MW00838244

HORSE ANATOMY COLORING BOOK

SCAN THE CODE TO ACCESS YOUR FREE DIGITAL COPY OF THE VETERINARY ANATOMY COLORING BOOK

The Veterinary Anatomy Coloring Book features:

- **The most effective way to skyrocket your anatomical knowledge of animals, all while having fun!**
- Full coverage of the major animal body systems to provide context and reinforce visual recognition
- **25 unique, easy-to-color illustrations of different animals with their anatomical terminology**
- Large 8.5 by 11-inch single side paper so you can easily remove your coloring
- **Self-quizzing for each illustration, with convenient same-page answer keys**

THIS BOOK BELONGS TO

TABLE OF CONTENTS

SECTION 1: THE SKELETON OF THE HORSE LATERAL ASPECT

SECTION 1: THE SKELETON OF THE HORSE LATERAL ASPECT

1. SKULL
2. ATLAS
3. BARS
4. AXIS
5. JAW
6. NECK VERTEBRAE
8. LUMBOSACRAL JOINT
7. LUMBAR VERTEBRA
9. POINT OF HIP
10. SACRUM
11. PELVIS
12. HIP JOINT
13. FEMUR
14. PATELLA
15. TIBIA
16. HOCK
17. STERNUM
18. ELBOW JOINT
19. RADIUS
20. KNEE
21. CANNON
22. SCAPULA
23. RIB CAGE
24. HUMERUS

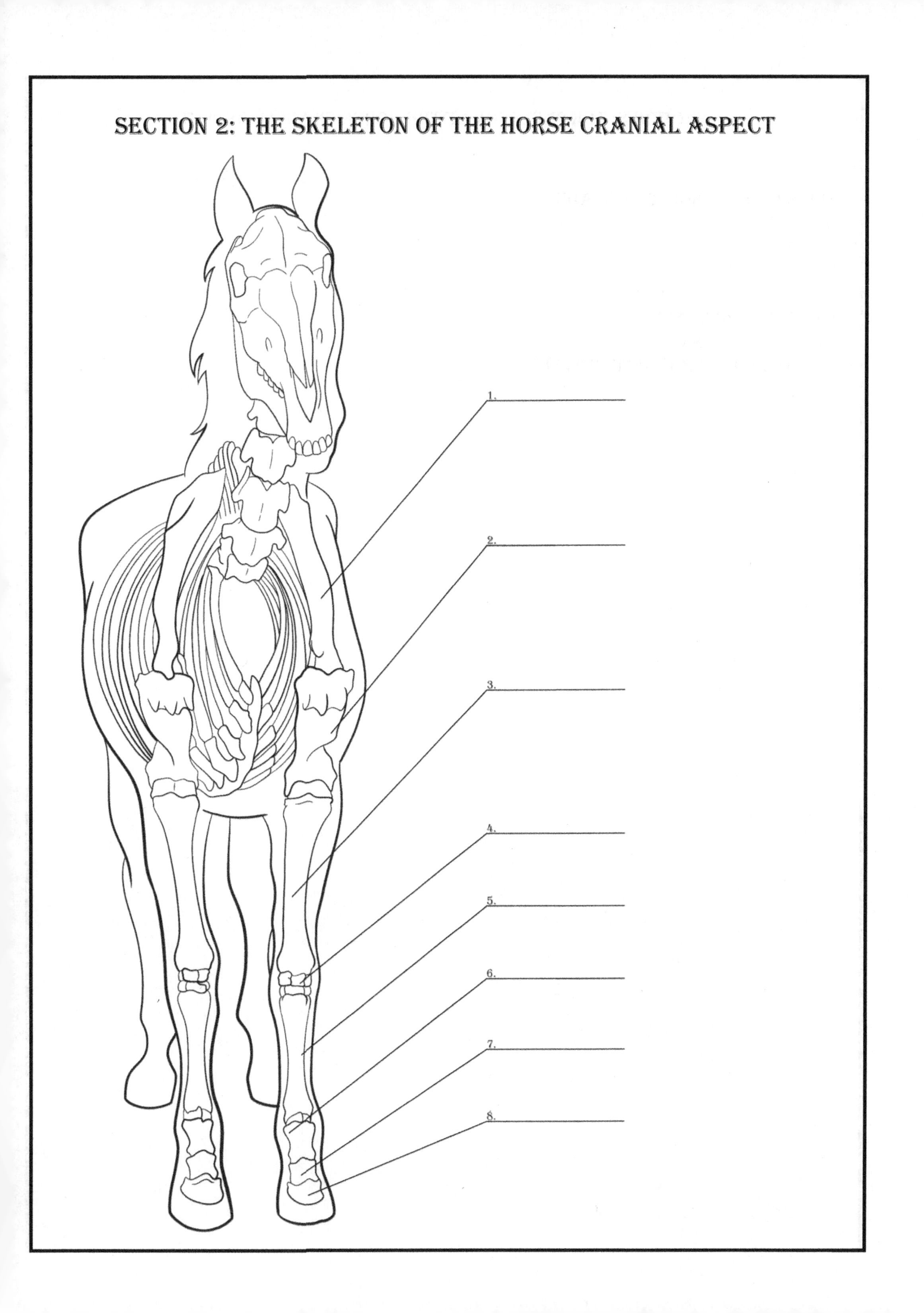
SECTION 2: THE SKELETON OF THE HORSE CRANIAL ASPECT
1.
2.
3.
4.
5.
6.
7.
8.

SECTION 2: THE SKELETON OF THE HORSE CRANIAL ASPECT

1. SPINE OF THE SHOULDER BLADE
2. HUMERUS
3. RADIUS
4. CARPAL BONES
5. 3RD METACARPAL BONE
6. PROXIMAL PHALANX
7. MIDDLE PHALANX
8. DISTAL PHALANX (COFFIN BONE)

SECTION 3: THE SKELETON OF THE HORSE CRANIAL AND CAUDAL ASPECT

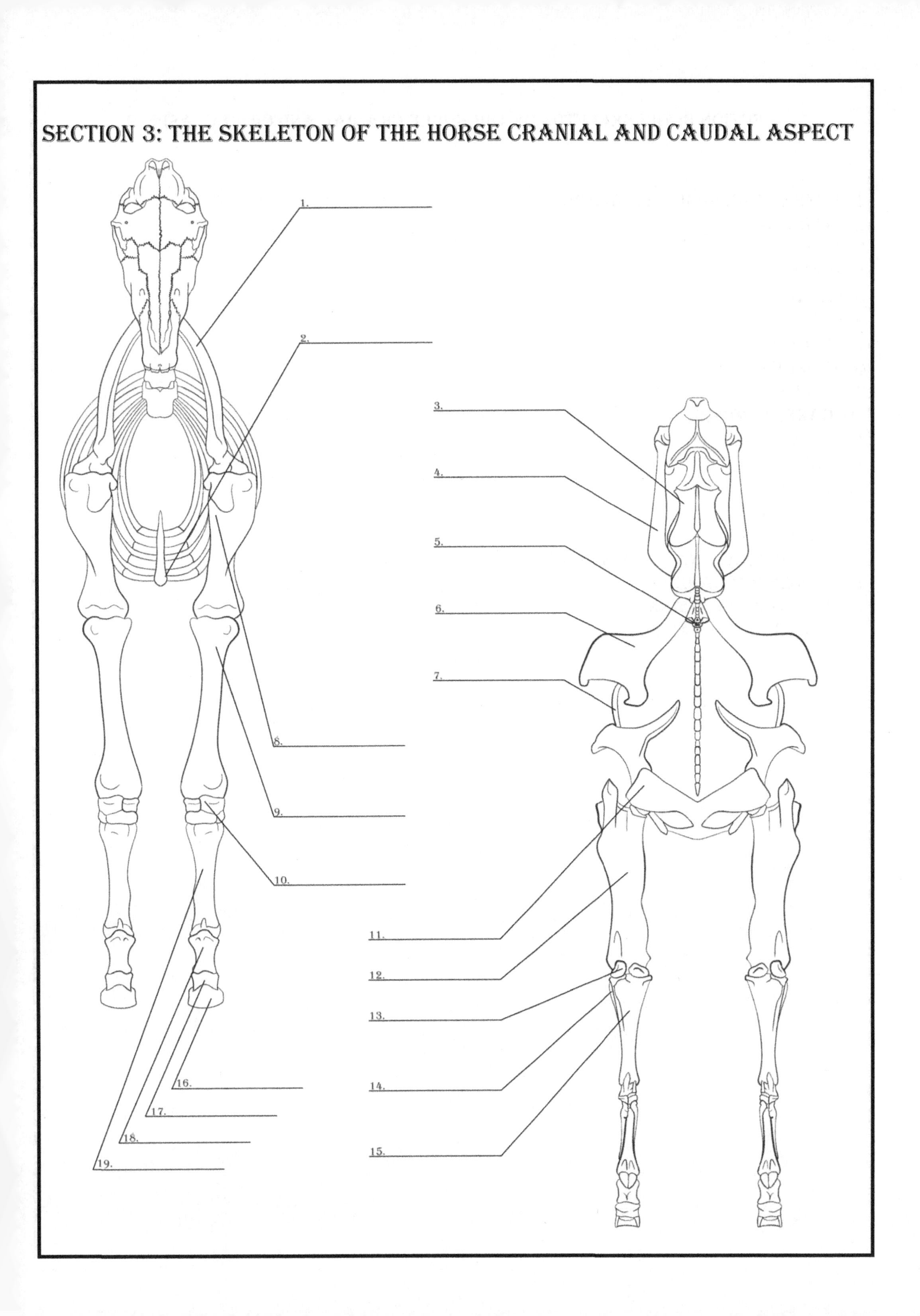

SECTION 3: THE SKELETON OF THE HORSE CRANIAL AND CAUDAL ASPECT

1. SPINE OF THE SHOULDER BLADE
2. STERNUM
3. AXIS
4. SKULL
5. SACRUM
6. SCAPULA
7. RIB CAGE
8. HUMERUS
9. RADIUS
10. CARPAL BONES
11. PELVIS
12. FEMUR
13. TALUS
14. SPLINT BONE
15. TIBIA
16. DISTAL PHALANX (COFFIN BONE)
17. MIDDLE PHALANX
18. PROXIMAL PHALANX
19. 3RD METACARPAL BONE

SECTION 4: THE SKELETON OF THE HORSE DORSAL ASPECT

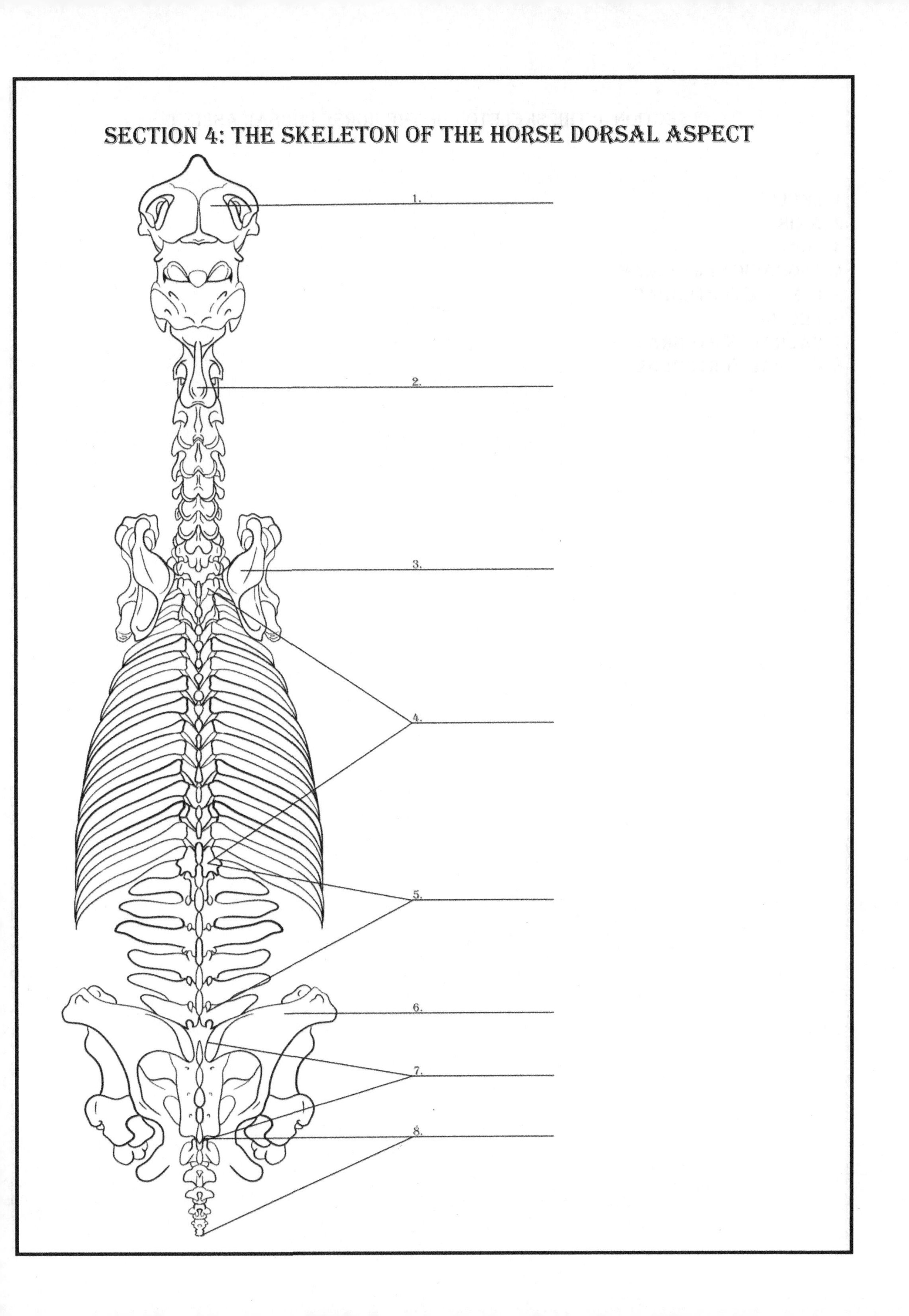

SECTION 4: THE SKELETON OF THE HORSE DORSAL ASPECT

1. SKULL
2. AXIS
3. SCAPULA
4. THORACIC VERTEBRAE
5. LUMBAR VERTEBRAE
6. PELVIS
7. SACRAL VERTEBRAE
8. CAUDAL VERTEBRAE

SECTION 5: THE MUSCLES OF THE HORSE LATERAL ASPECT

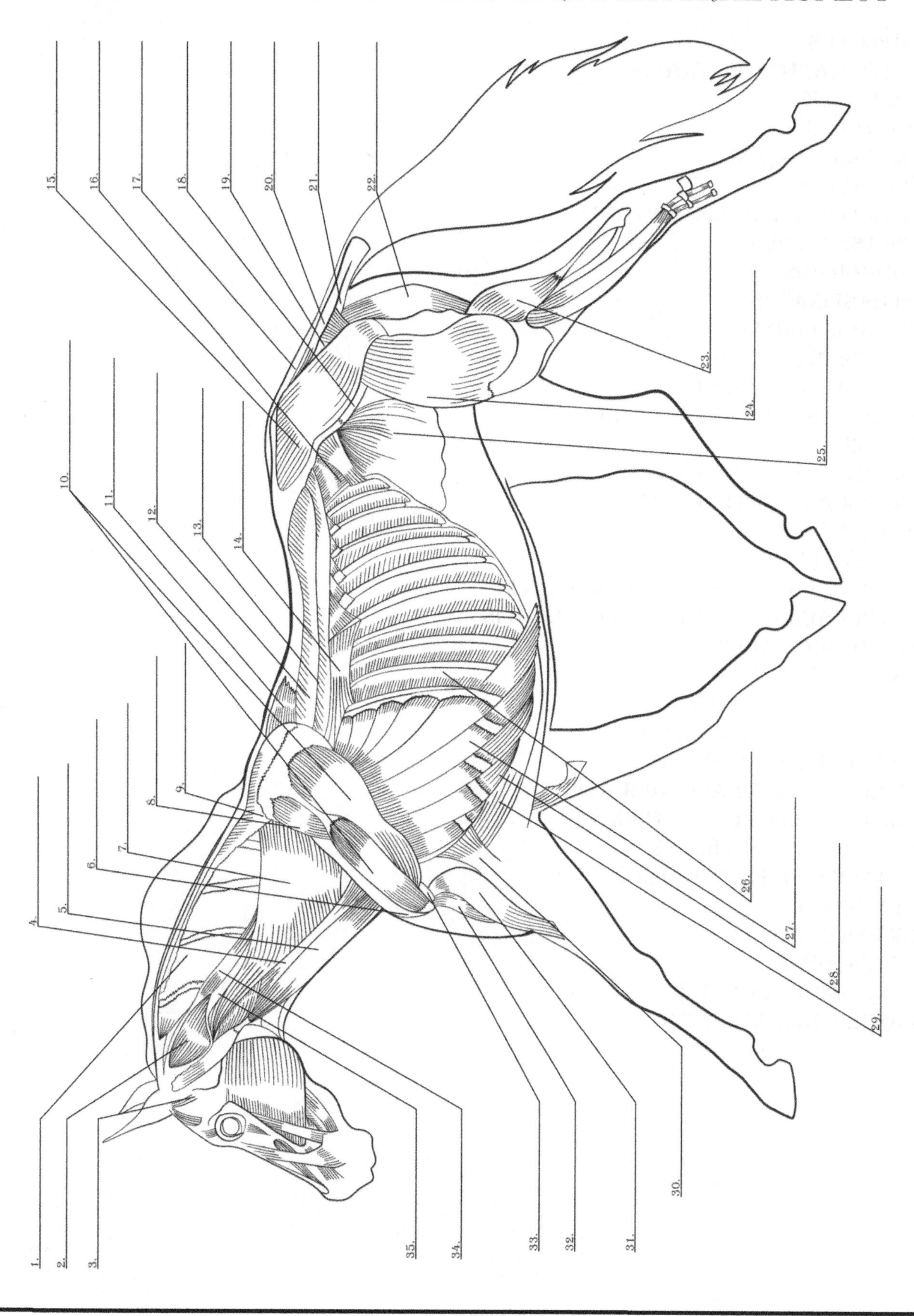

SECTION 5: THE MUSCLES OF THE HORSE LATERAL ASPECT

1. COMPLEXUS
2. RECTUS WAPITIS VENTRALIS
3. TEMPORALIS
4. OMOHYOIDEUS
5. STERNOCEPHALICUS
6. SUBCLAVIAN
7. SERRATUS VENTRALIS CERVICIS
8. SUPRASPINATIUS
9. RHOMBOIDEUS
10. INFRASPINATUS
11. SPINALIS DORSI
12. LONGISSIMUS DORSI
13. LONGISSIMUS COSTARUM
14. SERRATUS DORSALIS POSTERIOR
15. GLUTEUS MEDIUS
16. TRANSVERSUS ABDOMINIS
17. SACROCAUDALIS DORSALIS MEDIUS
18. ILIACUS
19. COCCYGEUS
20. SACROCAUDALIS DORSALIS LATERALIS
21. SACROCAUDALIS VENTRALIS LATERALIS
22. SEMIMEMBRANOUS
23. GASTRONEMIUS
24. QUADRICEPS FEMORIS
25. OBLIQUE ABDOMINIS INTERNUS
26. EXTERNAL INTERCOSTAL
27. SERRATUS VENTRALIS THORACIS
28. OBLIQUE ABDOMINIS EXTERNUS
29. PECTORALIS ASCENDENS
30. TRANSVERSE PECTORALIS
31. BRACHIALIS
32. BICEPS BRACHII
33. TERES MINOR
34. LONGISSIMUS CAPYTUS
35. LONGISSIMUS ATLANTIS

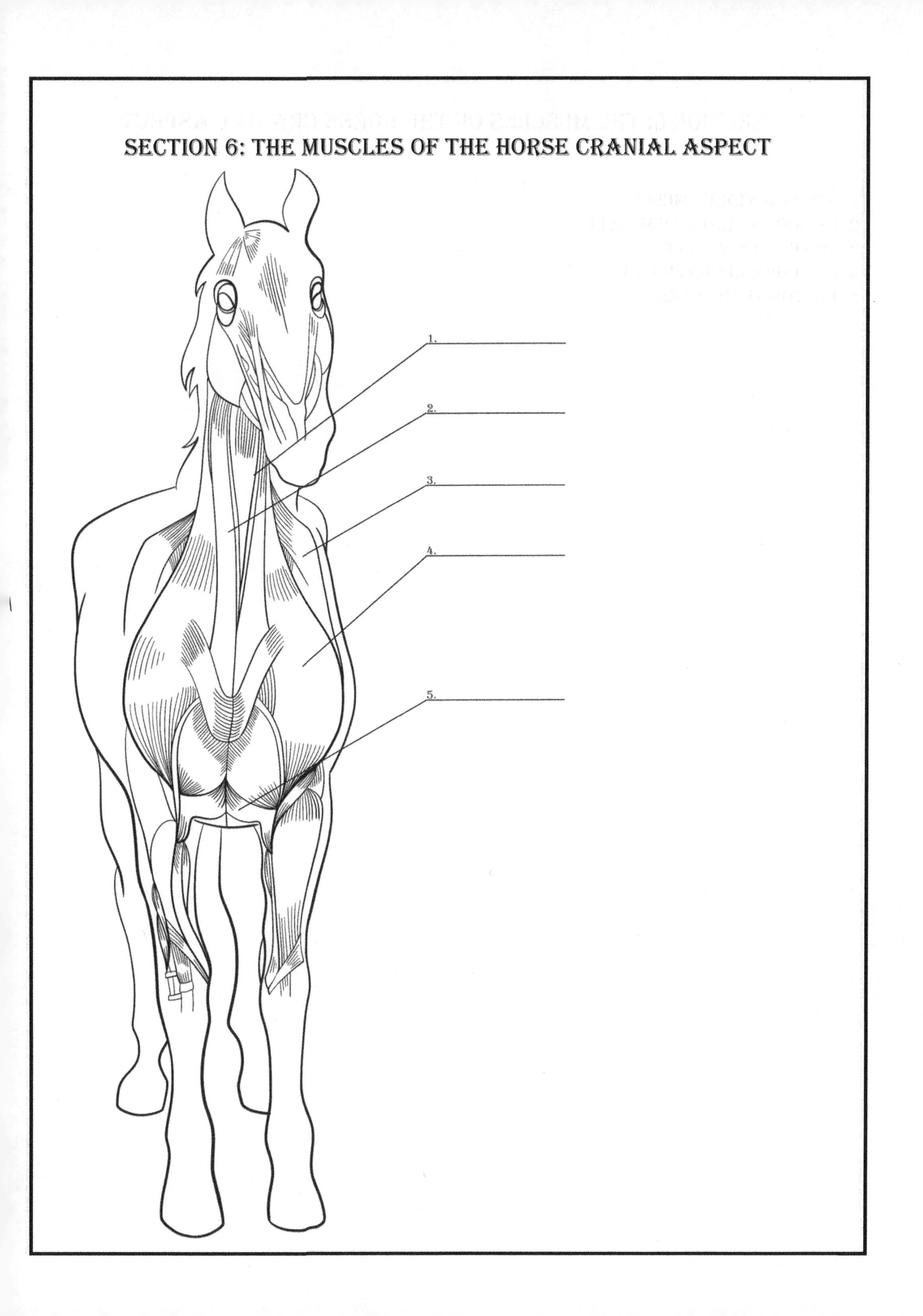
SECTION 6: THE MUSCLES OF THE HORSE CRANIAL ASPECT
1.
2.
3.
4.
5.

SECTION 6: THE MUSCLES OF THE HORSE CRANIAL ASPECT

1. STERNOHYIDEUS MUSCLE
2. STERNOCEPHALICUS MUSCLE
3. TRAPEZIUS MUSCLE
4. BRACHIOCEPUHALICUS MUSCLE
5. PECTORAL MUSCLES

SECTION 7: THE MUSCLES OF THE HORSE CRANIAL AND CAUDAL ASPECT

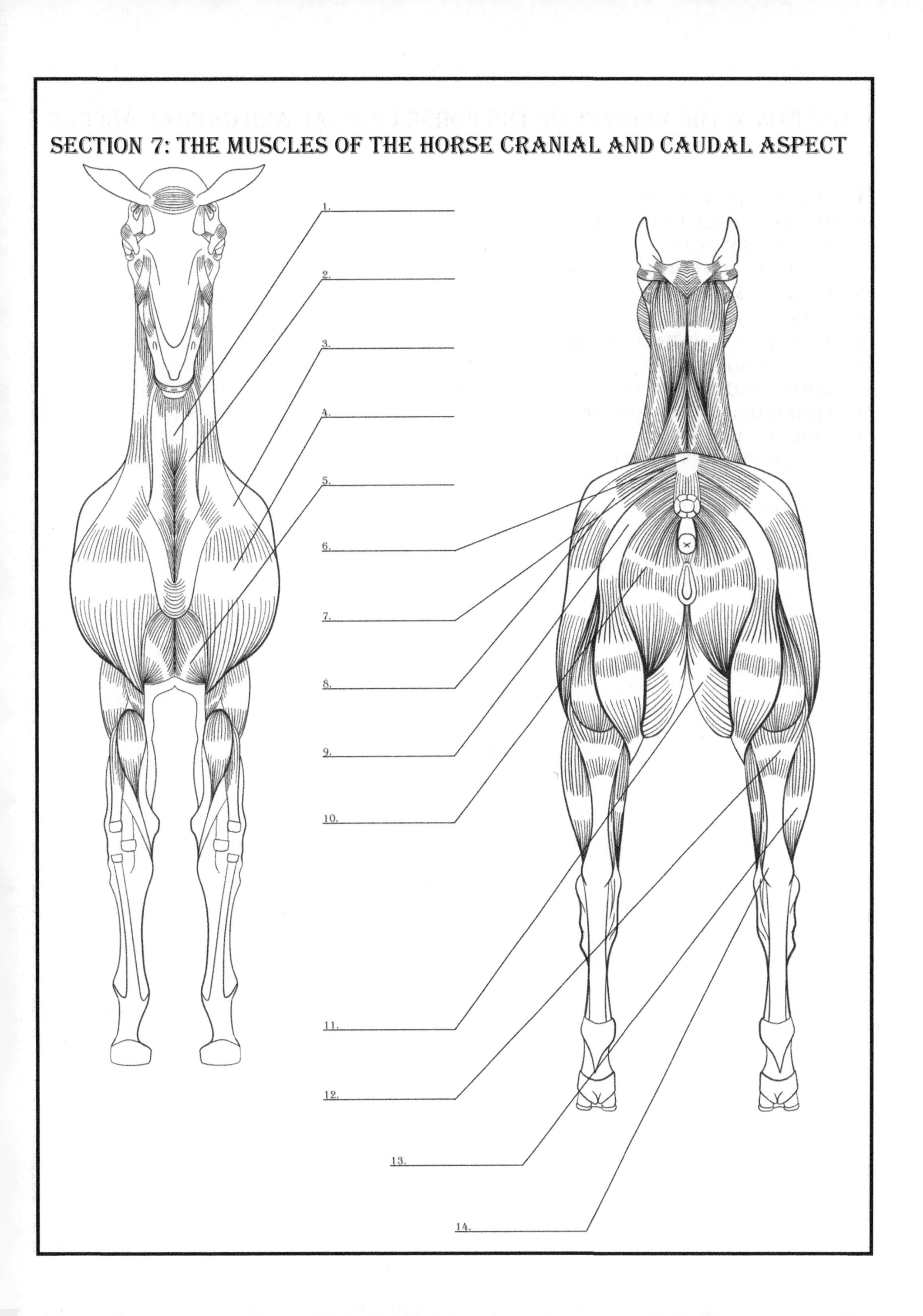

SECTION 7: THE MUSCLES OF THE HORSE CRANIAL AND CAUDAL ASPECT

1. STERNOHYIDEUS MUSCLE
2. STERNOCEPHALICUS MUSCLE
3. TRAPEZIUS MUSCLE
4. BRACHIOCEPUHALICUS MUSCLE
5. PECTORAL MUSCLES
6. TUBER SACRALE
7. GLUTEUS SUPERFICIALIS MUSCLE
8. BICEPS FEMORIS MUSCLE
9. SEMITENDINOSUS MUSCLE
10. SBMIMEMBRANOSUS MUSCLE
11. GRACILIS MUSCLE
12. GASTROCNEMIUS MUSCLE
13. TIBIALIS CRANIALIS MUSCLE
14. ACHILLES TENDON

SECTION 8: THE MUSCLES OF THE HORSE VENTRAL ASPECT

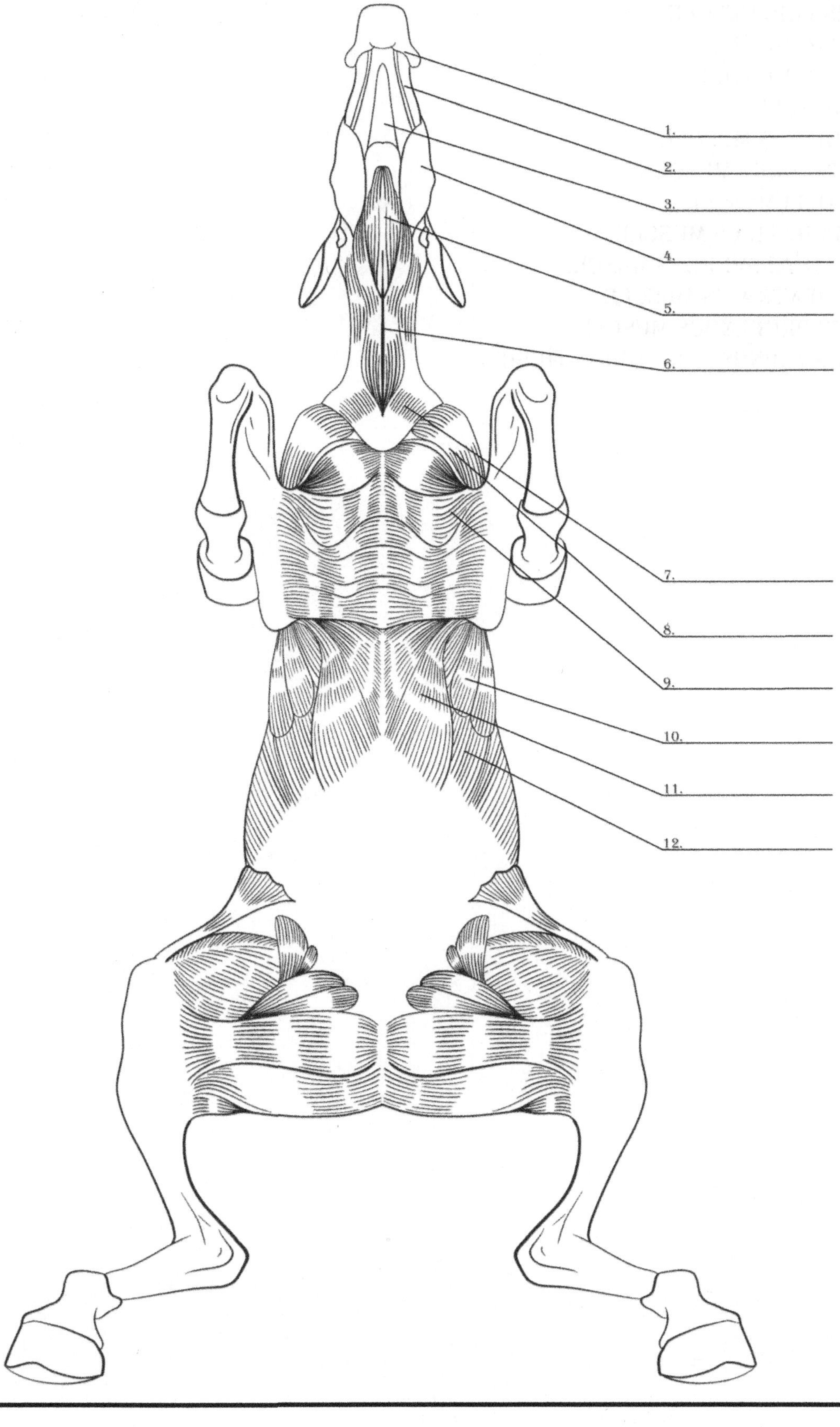

1. ____________________

2. ____________________

3. ____________________

4. ____________________

5. ____________________

6. ____________________

7. ____________________

8. ____________________

9. ____________________

10. ____________________

11. ____________________

12. ____________________

SECTION 8: THE MUSCLES OF THE HORSE VENTRAL ASPECT

1. ORBICULARIS FRIS MUSCLE
2. BUCCINATOR MUSCLE
3. MYLOHYOIDEUS MUSCLE
4. MASSETER MUSCLE
5. STERNOHPYOIDEUS MUSCLE
6. STERNOMASTOIDEUS MUSCLE
7. CUTANEUS COLLI MUSCLE
8. BRACHIOCEPUHALICUS MUSCLE
9. PECTORALIS TRANSVERSUS MUSCLE
10. SERRATUS VENTRALIS MUSCLE
11. PECTORALIS PROFUNDUS MUSCLE
12. OBLIQUUS EXTERNUS ABDOMINAL MUSCLE

SECTION 9: THE MUSCLES OF THE HORSE DORSAL ASPECT

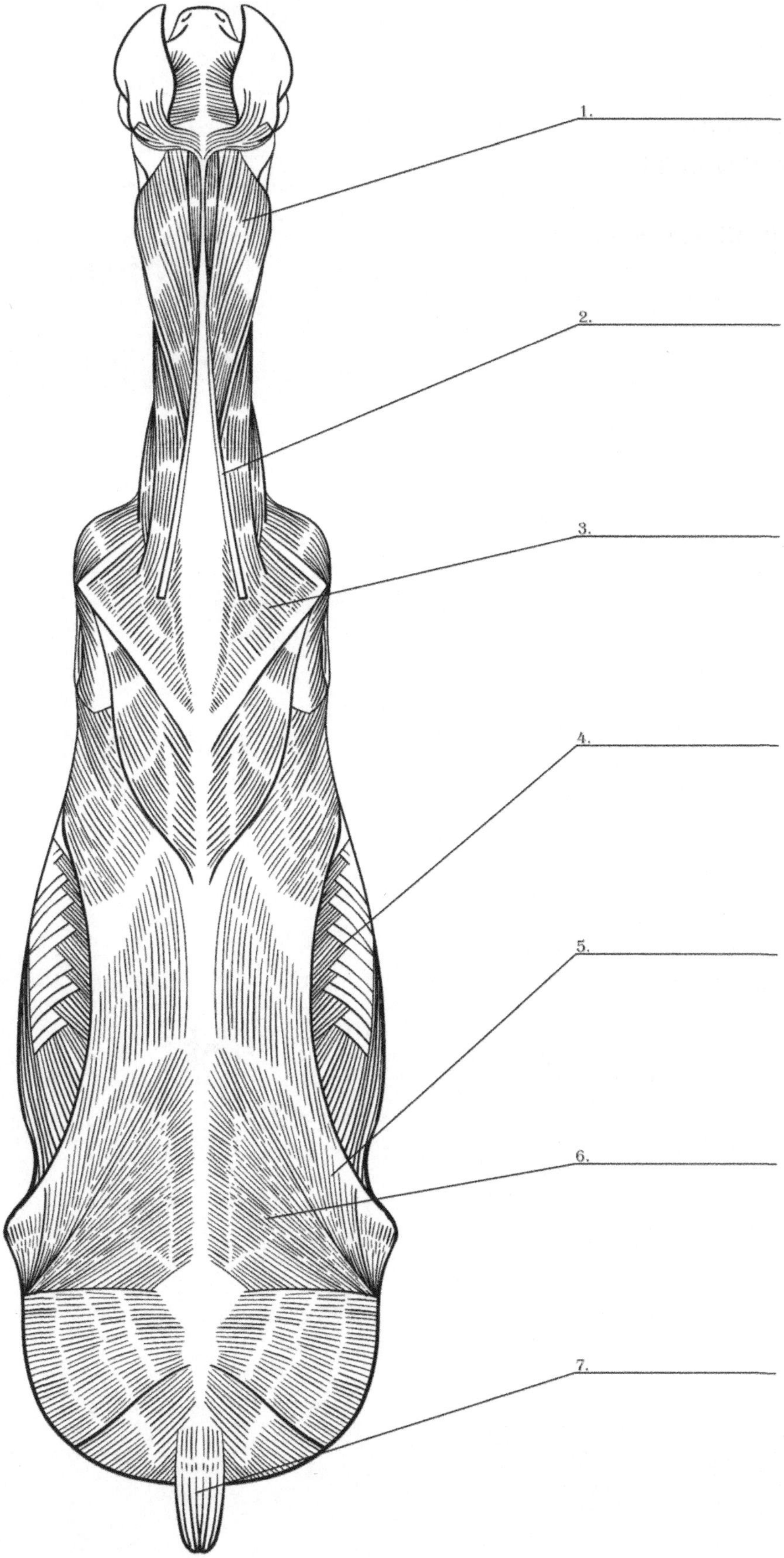

1. ____________

2. ____________

3. ____________

4. ____________

5. ____________

6. ____________

7. ____________

SECTION 9: THE MUSCLES OF THE HORSE DORSAL ASPECT

1. COMPLEXUS MUSCLE
2. RHOMBOIDEUS MUSCLE
3. SPINALIS DORSI MUSCLE
4. EXTERNAL INTERCOSTAL MUSCLE
5. OBLIQUE ABDOMINIS INTERNUS MUSCLE
6. GLUTEUS MEDIUS MUSCLE
7. SACROCAUDALIS DORALIS MEDIUS MUSCLE

SECTION 10: INTERNAL ORGANS OF THE HORSE

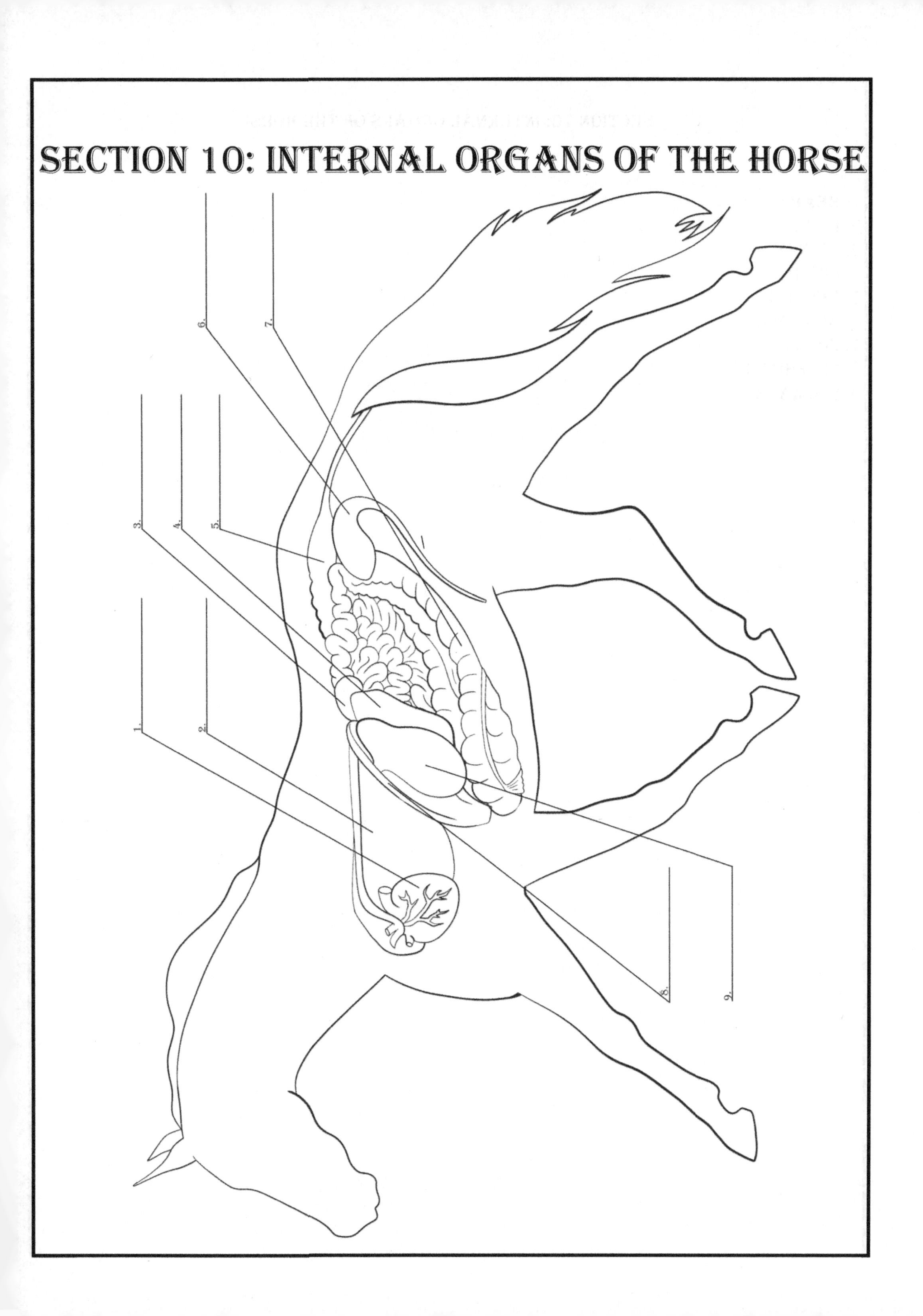

SECTION 10: INTERNAL ORGANS OF THE HORSE

1. HEART
2. LUNG
3. KIDNEY
4. LIVER
5. RECTUM
6. BLADDER
7. COLON
8. DIAPHRAGM
9. STOMACH

SECTION 11: BLOOD VESSELS OF THE HORSE

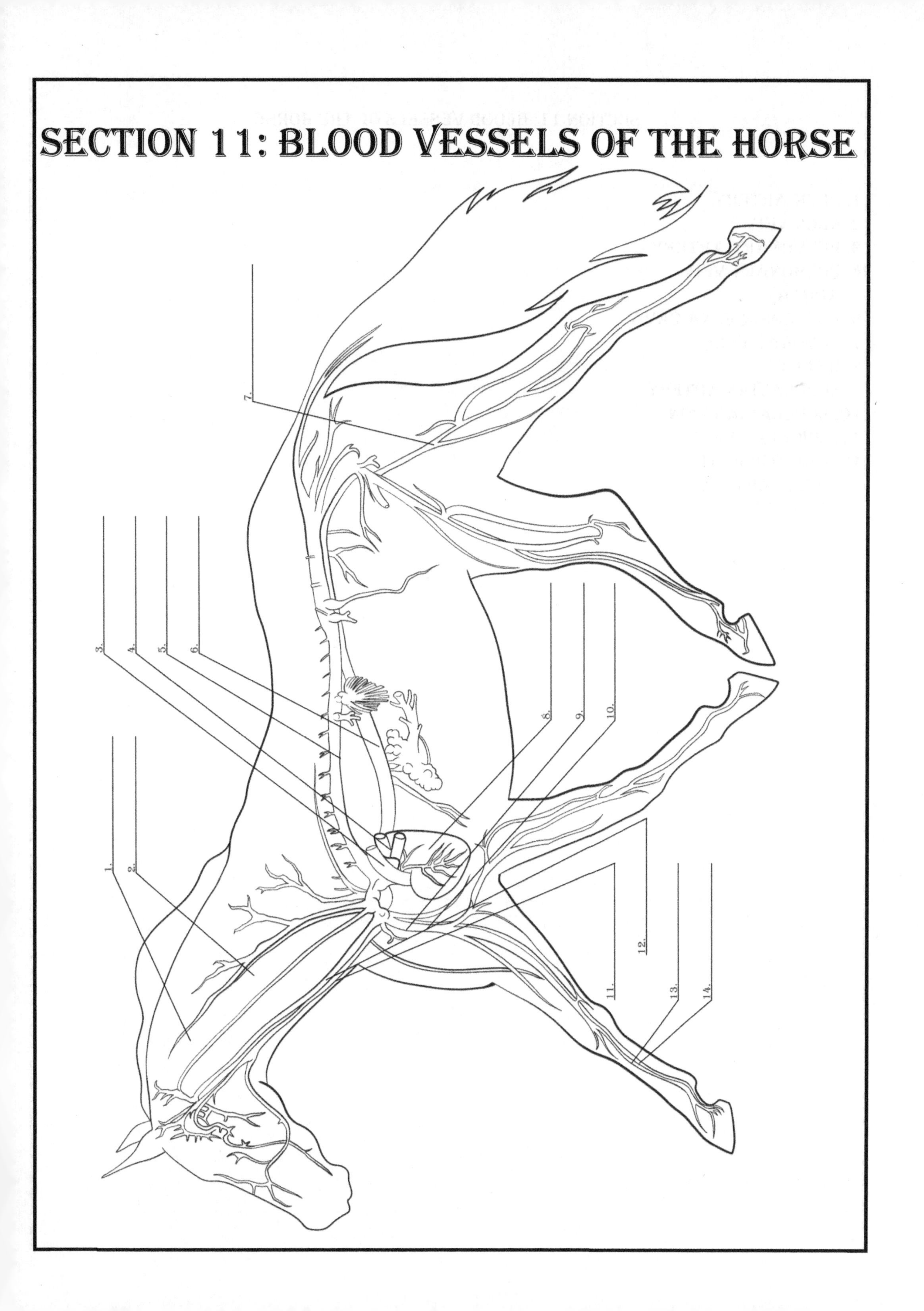

SECTION 11: BLOOD VESSELS OF THE HORSE

1. NECK ARTERY
2. NECK VEIN
3. PULMONARY ARTERY
4. PULMONARY VEIN
5. AORTA
6. POSTERIOR VENA CAVA
7. FEMORAL VEIN
8. HEART
9. SUBCLAVIAN ARTERY
10. SUBCLAVIAN VEIN
11. JUGULAR VEIN
12. CAROTID ARTERY
13. PEDIS ARTERY
14. PEDIS VEIN

SECTION 12: NERVES OF THE HORSE

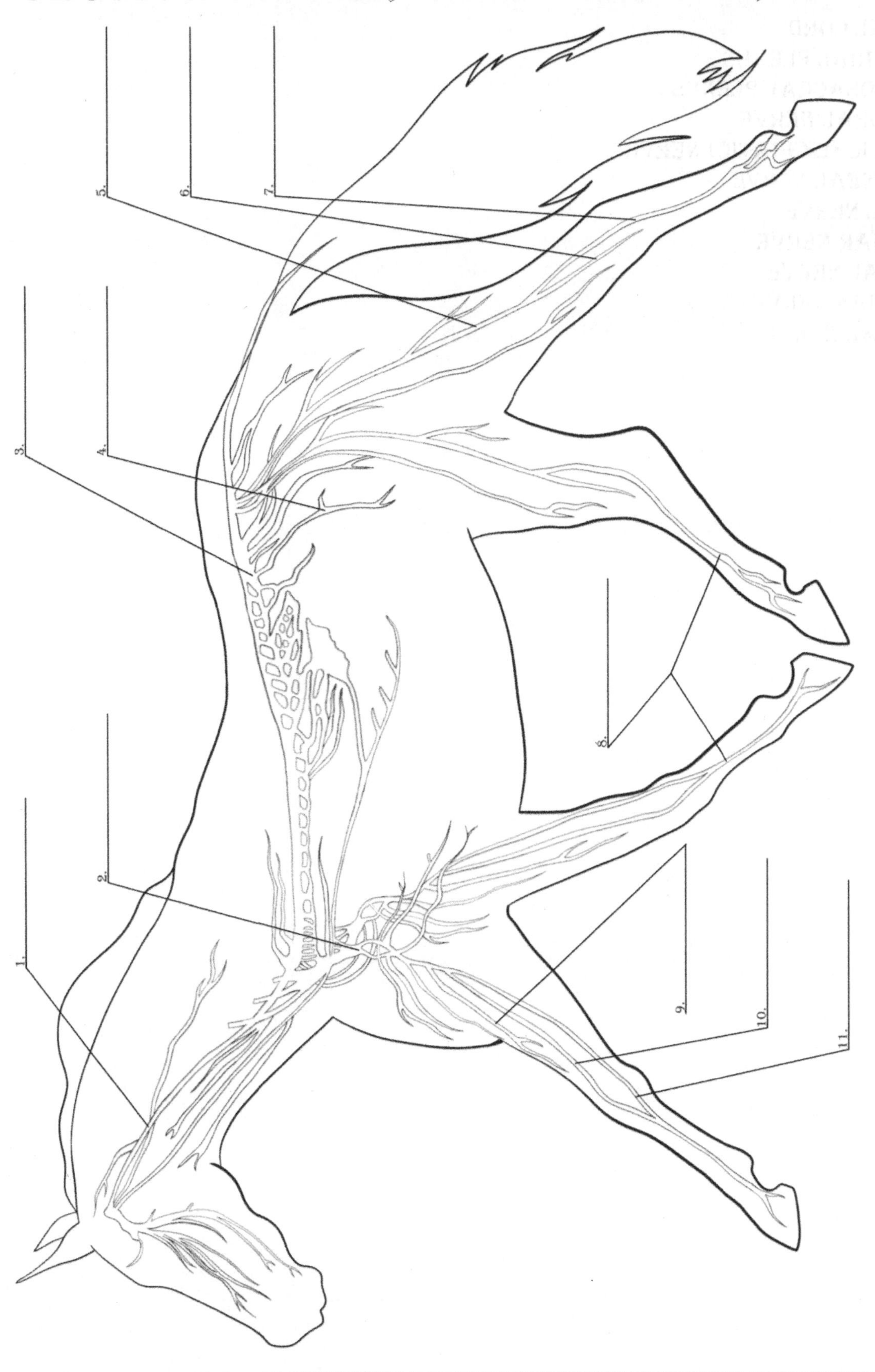

SECTION 12: NERVES OF THE HORSE

1. SPINAL CORD
2. BRACHIAL PLEXUS
3. LUMBOSACRAL PLEXUS
4. FEMORAL NERVE
5. SCIATIC (ISCHIATIC) NERVE
6. PERONEAL NERVE
7. TIBIAL NERVE
8. PALMAR NERVE
9. RADIAL NERVE
10. MEDIAN NERVE
11. ULNAR NERVE

SECTION 13: THE SKULL OF THE HORSE LATERAL ASPECT

SECTION 13: THE SKULL OF THE HORSE LATERAL ASPECT

1. INCISIVAL BONE
2. NASAL BONE
3. INFRAORBITAL HOLE
4. MAXILLA
5.LACRIMAL BONE WITH THE ORBIT BEHIND IT
6. FRONTAL BONE
7. PARIETAL BONE
8. FOSSA TEMPORALIS
9. MEATUS ACUSTICUS EXTERNES
10. NUCHAL CREST
11. OCCIPITAL CONDYLE
12. PARACONDYLAR PROCESS
13. ZYGOMATIC ARCH
14. ZYGOMATIC BONE WITH FACIAL CREST
15. MANDIBULAR ANGLE
16. MOLAR TEETH
17. PREMOLAR TEETH
18. MARGO INTERALVEOLARIS
19. INSICORS
20. INCISOR TEETH

SECTION 14: INSIDE THE SKULL OF THE HORSE LATERAL ASPECT

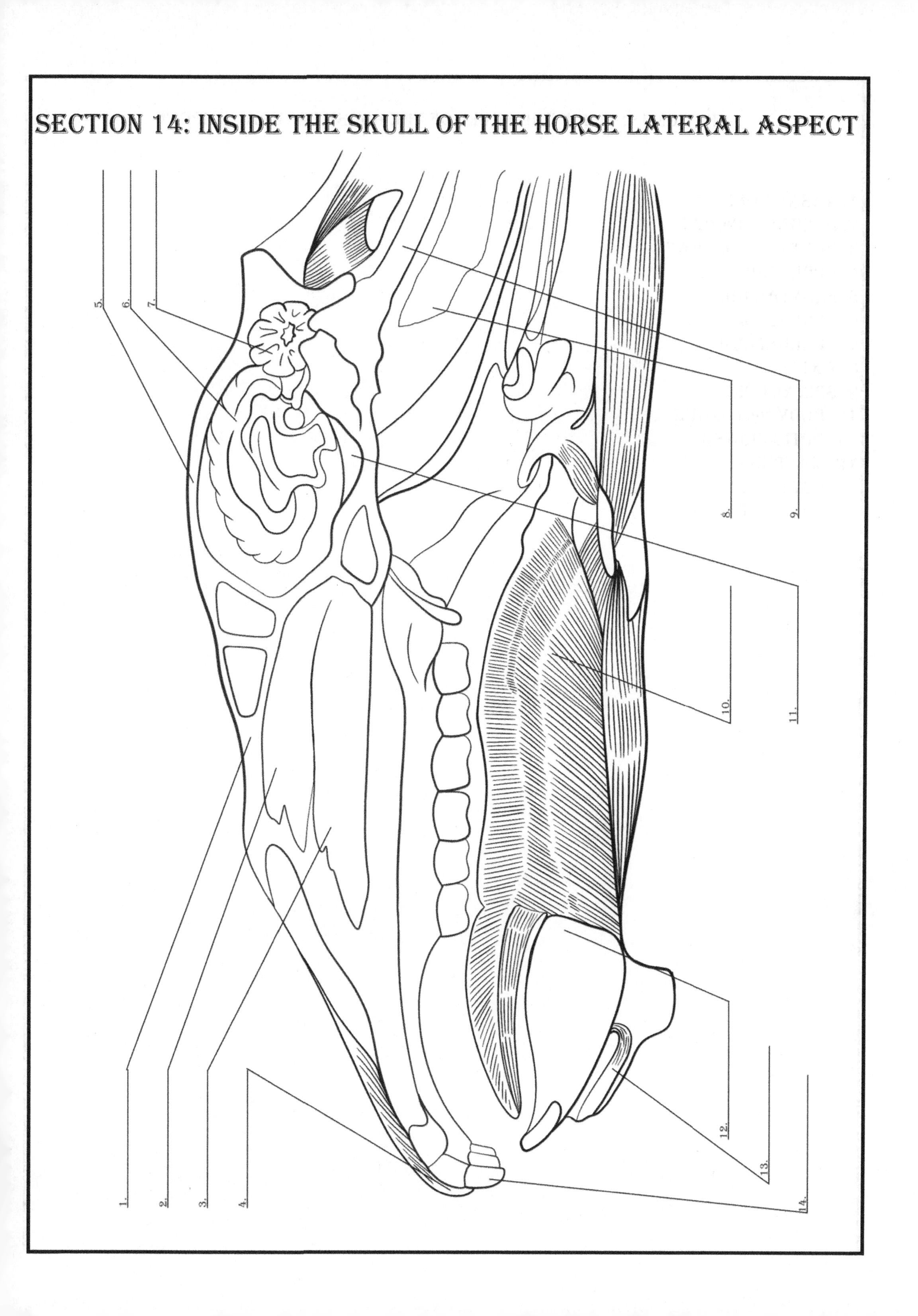

SECTION 14: INSIDE THE SKULL OF THE HORSE LATERAL ASPECT

1. NASAL BONE
2. DORSAL CONCHAE
3. VENTRAL CONCHAE
4. UPPER LIP
5. FRONTAL BONE
6. CEREBEUM
7. CEREBELLUM
8. AXIS
9. SPINAL CORD
10. BODY OF TONGUE
11. OPTIC CHIASM
12. MANDIBLE
13. LOWER LIP
14. INCISOR TEETH

SECTION 15: THE SKULL OF THE HORSE DORSAL ASPECT

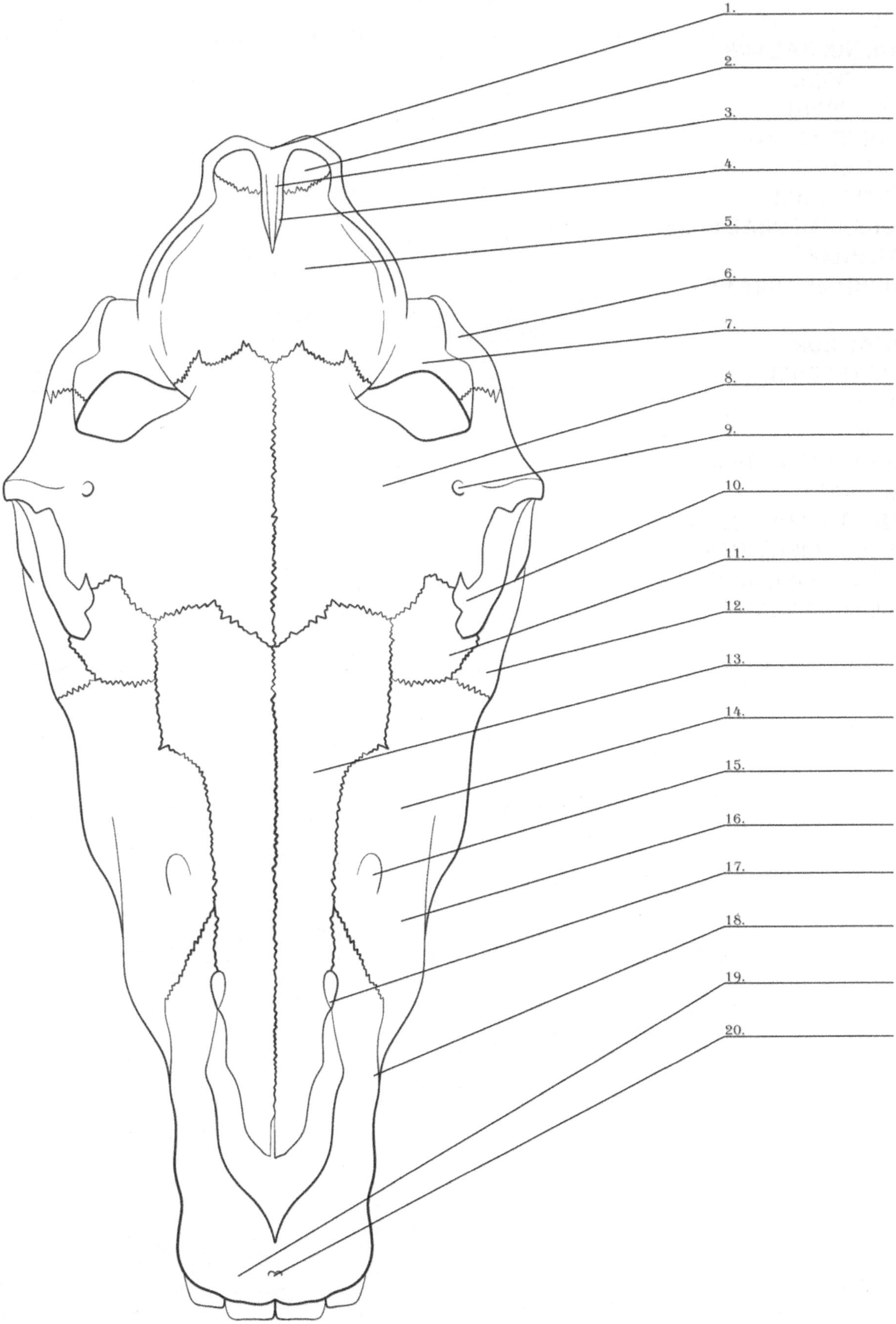

1. ____________________
2. ____________________
3. ____________________
4. ____________________
5. ____________________
6. ____________________
7. ____________________
8. ____________________
9. ____________________
10. ____________________
11. ____________________
12. ____________________
13. ____________________
14. ____________________
15. ____________________
16. ____________________
17. ____________________
18. ____________________
19. ____________________
20. ____________________

SECTION 15: THE SKULL OF THE HORSE DORSAL ASPECT

1. SUPERIOR NUCHAL LINE
2. OCCIPITAL BONE
3. PARIETAL CREST
4. INTERPARIETAL BONE
5. PARIETAL BONE
6. ZYGOMATIC ARCH
7. SQUAMOUS TEMPORAL BONE
8. FRONTAL BONE
9. SUPRAORBITAL FORAMEN
10. ORBIT
11. LACRIMAL BONE
12. ZYGOMATIC BONE
13. NASAL BONE
14. MAXILLA
15. INFRAORBITAL FORAMEN
16. FACIAL CREST
17. NASOMAXILLARY NOTCH
18. NASAL OF INCISIVE BONE
19. BODY OF INCISIVE BONE
20. INCISIVE FORAMEN

SECTION 16: THE SKULL OF THE HORSE VENTRAL ASPECT

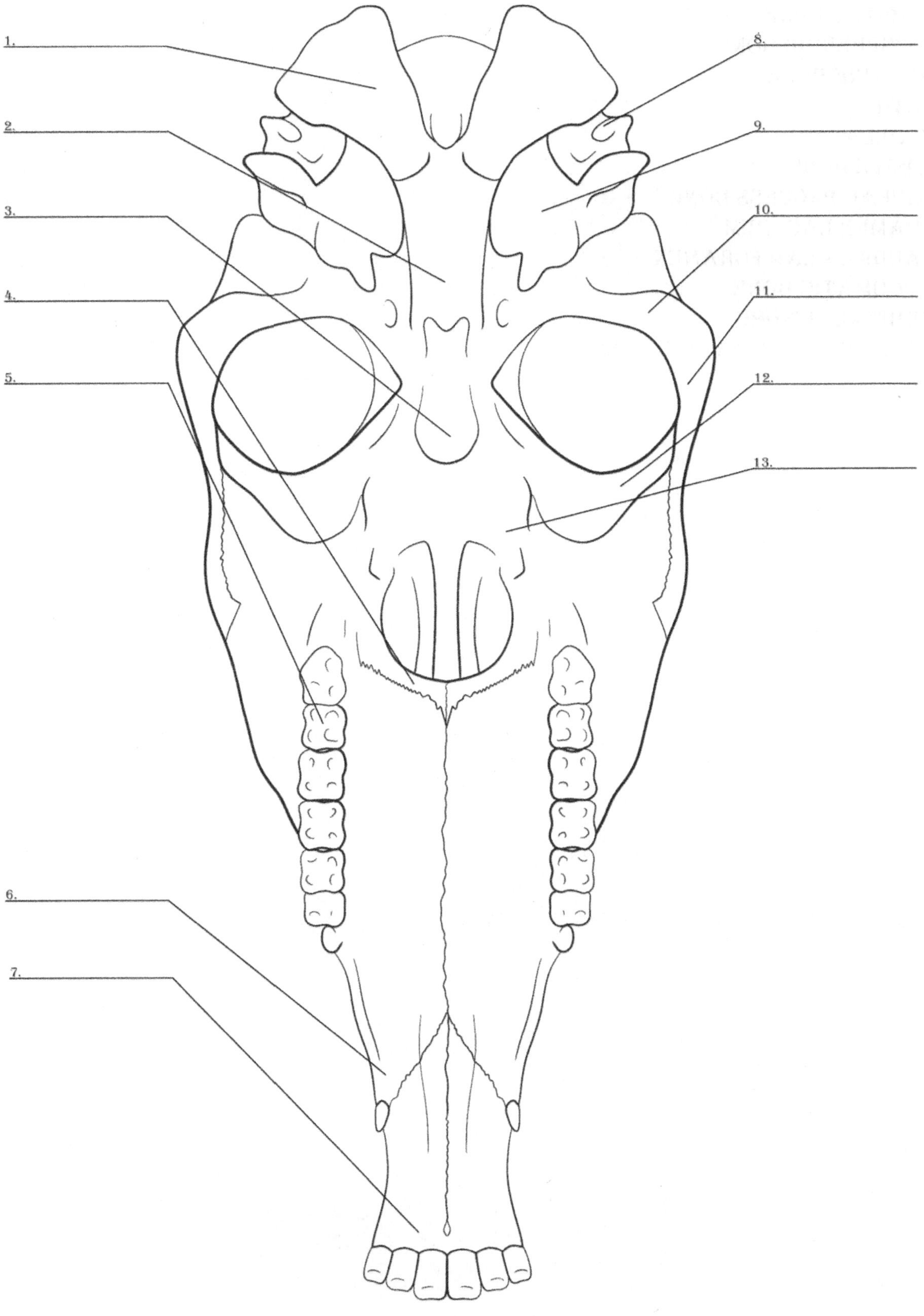

1. ____________
2. ____________
3. ____________
4. ____________
5. ____________
6. ____________
7. ____________
8. ____________
9. ____________
10. ____________
11. ____________
12. ____________
13. ____________

SECTION 16: THE SKULL OF THE HORSE VENTRAL ASPECT

1. FORAMEN MAGNUM
2. OCCIPITAL BONE
3. BASISPHENOID BONE
4. PALATINE BONE
5. TEETH
6. MAXILLA
7. INCISIVE BONE
8. JUGULAR PROCESS BONE
9. FORAMEN LACERUM
10. CAUDAL ALAR FORAMEN
11. ZYGOMATIC BONE
12. ORBITAL FISSURE
13. HAMULUS OF PTERYGOID BONE

SECTION 17: THE MUSCLES OF THE HEAD LATERAL ASPECT

SECTION 17: THE MUSCLES OF THE HEAD LATERAL ASPECT

1. CANINUS MUSCLE
2. LEVATOR LABII MAXILLARIS MUSCLE
3. LEVATOR NASOBIALIS MUSCLE
4. LEVATOR ANGULI MEDIALIS MUSCLE
5. INTERSCUTULARIS MUSCLES
6. PARS TEMPORALIS OF THE FRONTOSCUTULARIS MUSCLE
7. CERVICOAURICULARIS MUSCLE
8. PARTOIDEOAURICULARIS MUSCLE
9. MASSETER MUSCLE
10. DEPRESSOR LABII MANDIBULARIS MUSCLE
11. BUCCALIS MUSCLE
12. ZYGOMATICUS MUSCLE
13. ORBICULARIS FRIS MUSCLE

SECTION 18: THE MUSCLES OF THE HEAD DORSAL ASPECT

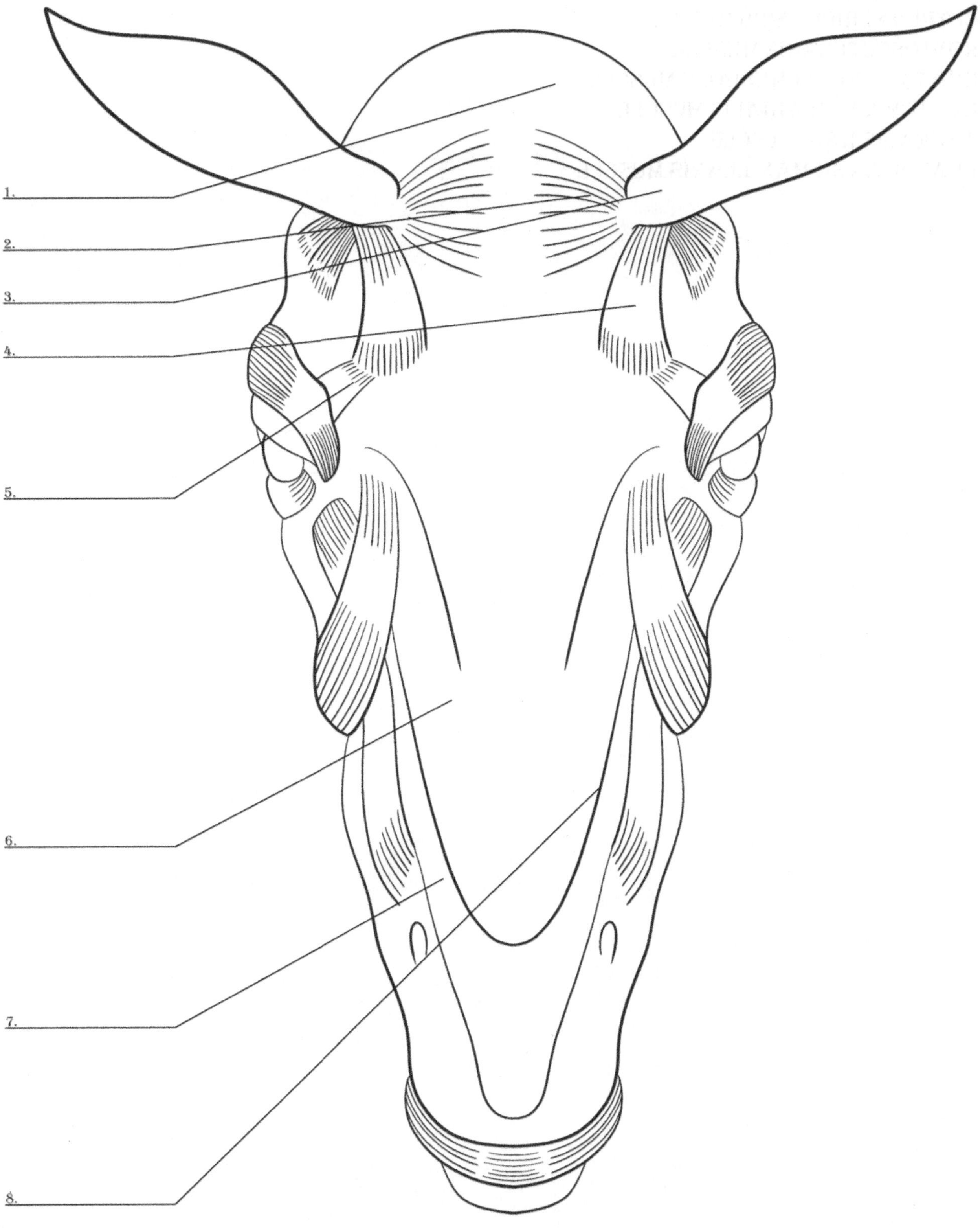

1. ____________

2. ____________

3. ____________

4. ____________

5. ____________

6. ____________

7. ____________

8. ____________

SECTION 18: THE MUSCLES OF THE HEAD DORSAL ASPECT

1. PERVICOAURICOLARIS SUPERFICIALIS MUSCLE
2. INTERSCUTULARIS MUSCLE
3. SCUTULOAURICULARIS MUSCLE
4. FRONTOSCUTULARIS MUSCLE
5. LEVATOR ANGULI MEDIALIS MUSCLE
6. LEVATOR NASOLABIALIS MUSCLE
7. LATERALIS NASI MUSCLE
8. LEVATOR LABII MAXILLARIS MUSCLE

SECTION 19: THE BRAIN OF THE HORSE LATERAL AND DORSAL ASPECT

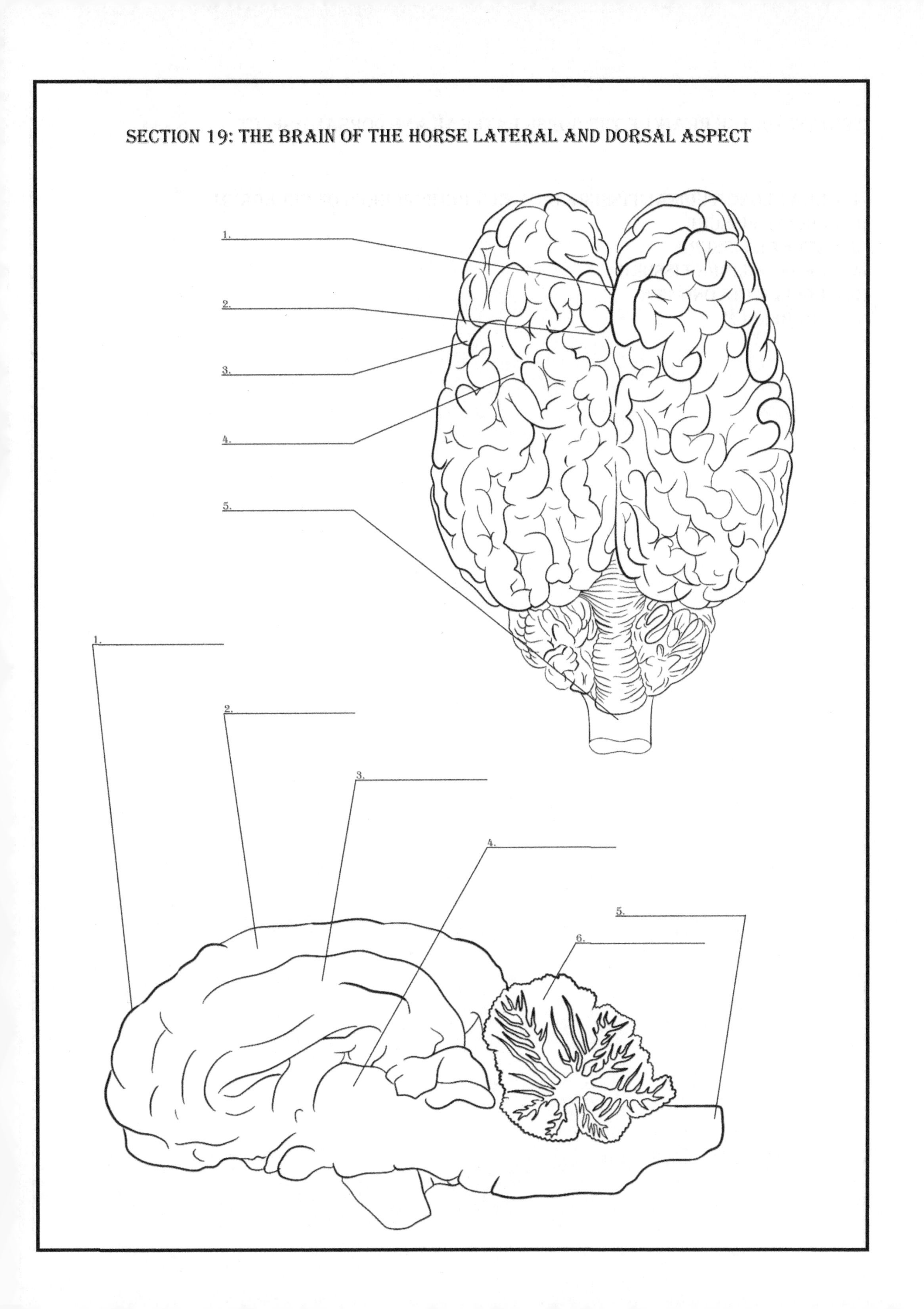

SECTION 19: THE BRAIN OF THE HORSE LATERAL AND DORSAL ASPECT

1. GREAT LONGITUDINAL FISSURE BETWEEN HEMISPHERES OF CEREBRUM
2. CRUCIAL FISSURE
3. LATERAL FISSURE
4. GREAT OBLIQUE FISSURE
5. MEDULLA OBLONGATA
6. CEREBELLUM

SECTION 20: THE EYE OF THE HORSE

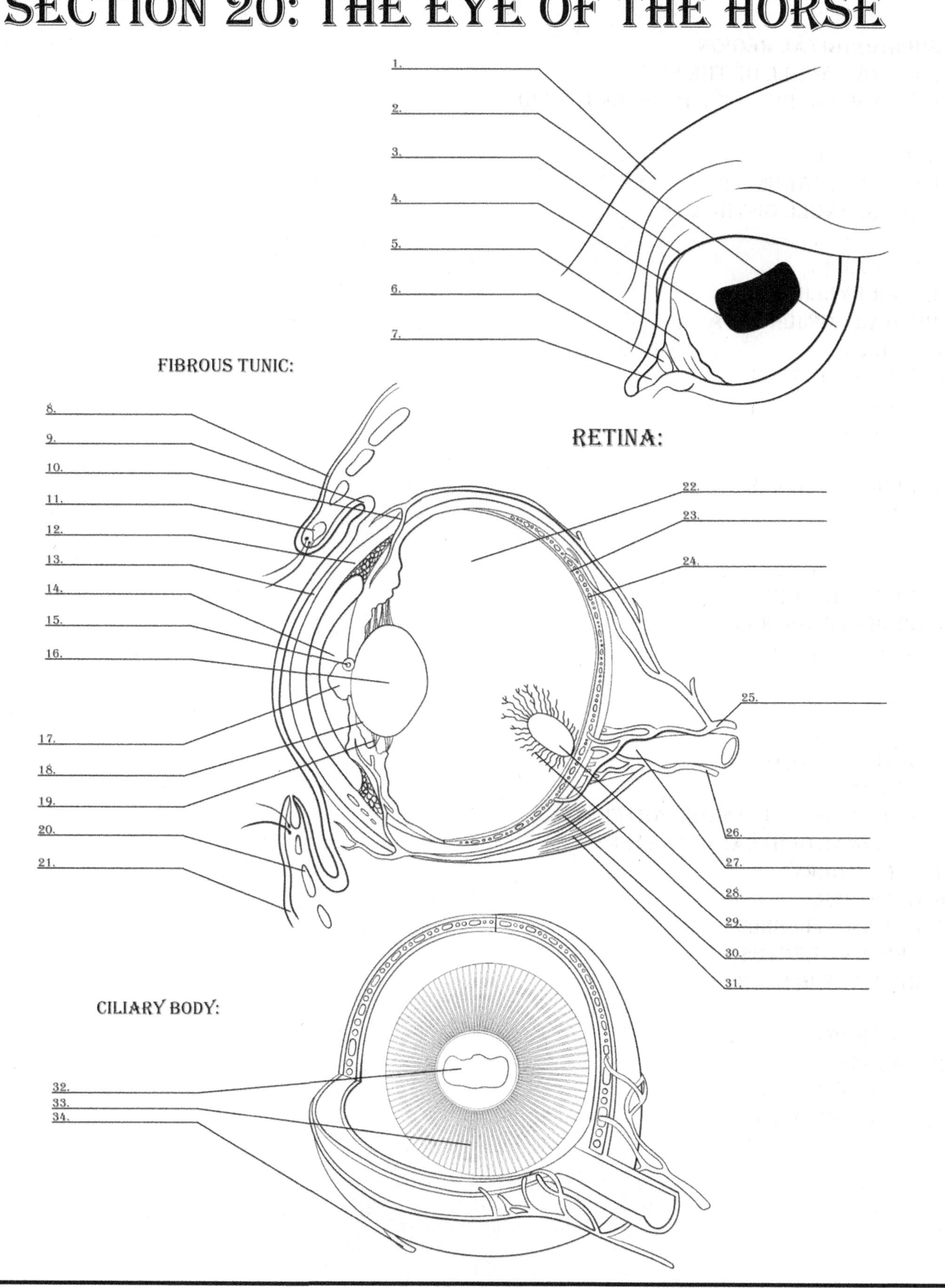

SECTION 20: THE EYE OF THE HORSE

1. SUPRAORBITAL REGION
2. LATERAL ANGLE OF THE EYE
3. EYELASH BORDER OF THE UPPER EYELID
4. IRIS
5. 3RD EYELID
6. LACRIMAL CARUNCLE
7. MEDIAL ANGLE OF THE EYE

FIBROUS TUNIC:
8. UPPER EYELID
9. BULBAR CONJUNCTIVA
10. SCLERA
11. TARSAL GLANDS
12. LIMBUS
13. CORNEA
14. IRIS
15. IRIDIC GRANULES
16. LENS
17. PUPIL
18. LENS CAPSULE
19. ZONULAR FIBERS
20. ORBICULARIS OCULI
21. LOWER EYELID

RETINA:
22. BLIND PART
23. OPTICAL PART
24. CHOROID
25. EXTERNAL OPHTHALMIC ARTERY
26. INTERNAL OPHTHALMIC ARTERY
27. OPRAC NERVE
28. OPTIC DISC
29. RETINAL VESSELS
30. VENTRAL RECTUS
31. RETRACTOR BULBI

CILIARY BODY:
32. RADII LENTIS
33. CILIARY CROWN
34. VORTICOSE VEINS

SECTION 21: THE LIPS AND NOSE OF THE HORSE

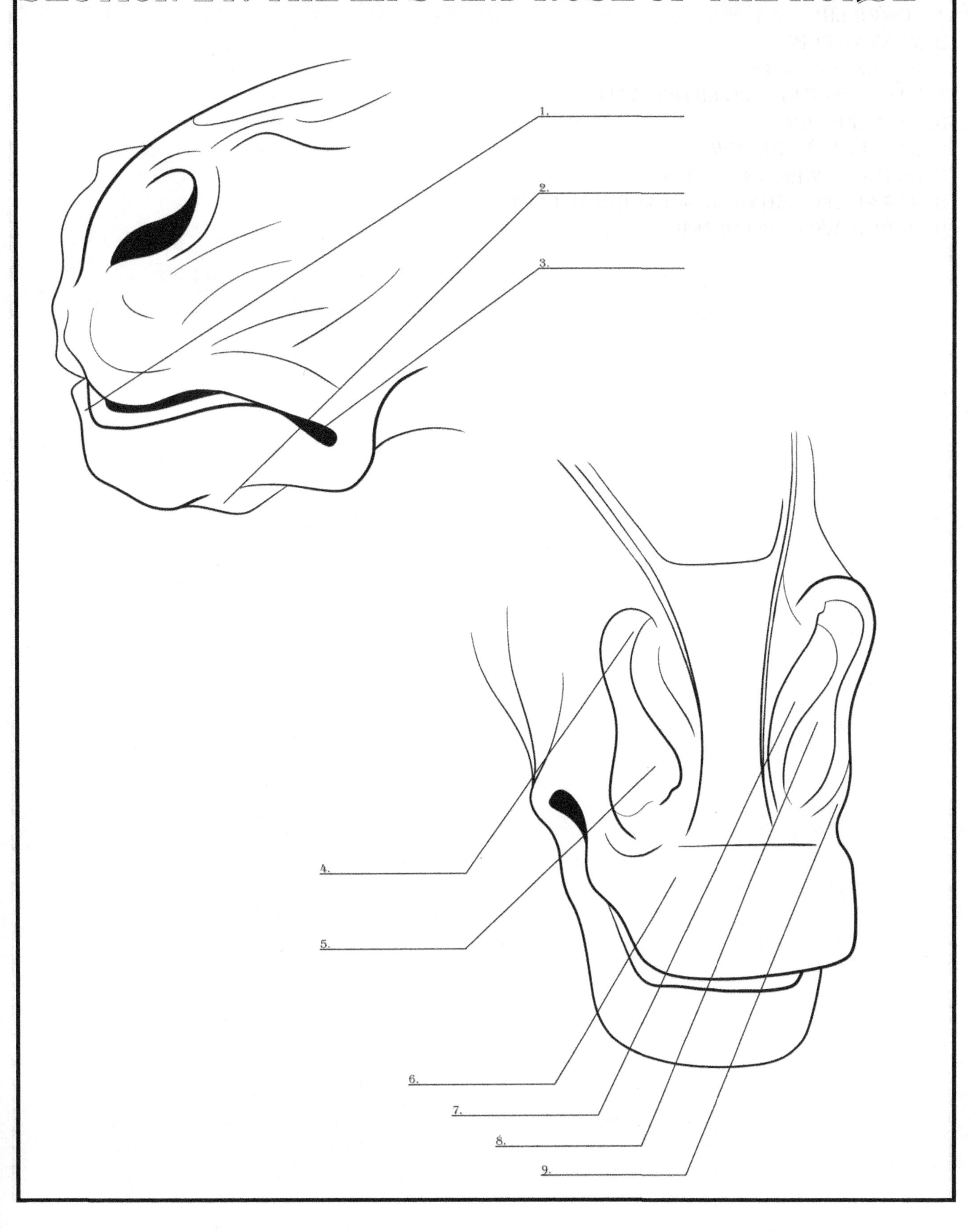

SECTION 21: THE LIPS AND NOSE OF THE HORSE

1. LOWER LIP
2. MENTAL POINT
3. ANGLE OF MOUTH
4. FALSE NOSTRIL (DIVERTICULUM)
5. TRUE NOSTRIL
6. NASOLABIAL REGION
7. LATERAL WING OF NOSTRILS
8. NASAL OPENING OF NASOLACRIMAL DUCT
9. MEDIAL WING OF NOSTRIL

SECTION 22: THE EARS OF THE HORSE

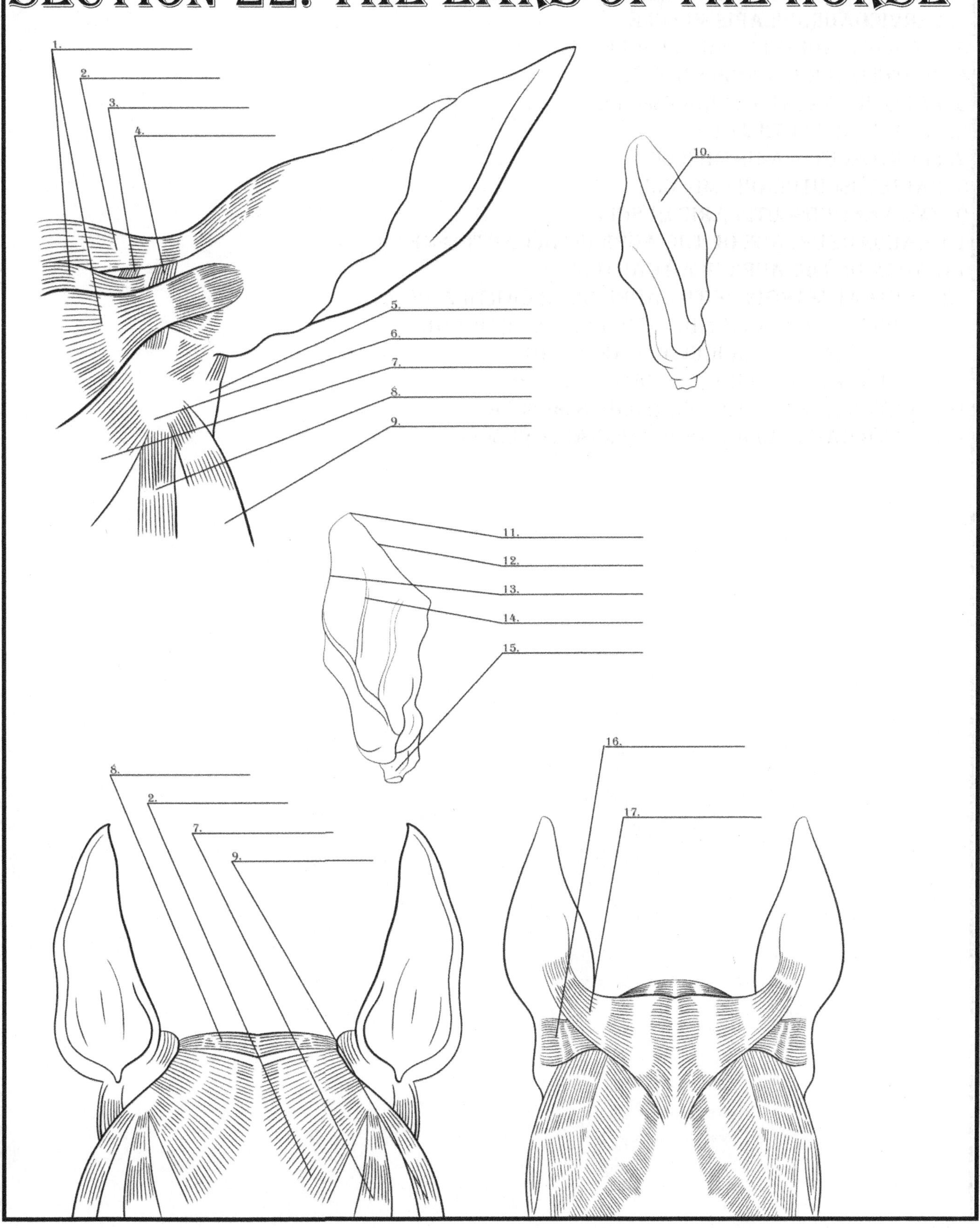

SECTION 22: THE EARS OF THE HORSE

1. INTER/PARIETOAURICULARIS MUSCLES
2. CERVICOAURICULARIS MUSCLE
3. ROTATOR AURIS LONGUS MUSCLE
4. SCUTULOAURICULARIS MUSCLE
5. PAROTIDEOAURICULARIS MUSCLE
6. SCUTULAR CARTILAGE
7. FRONTOSCUTULARIS MUSCLE
8. PARIETOSCUTULARIS MUSCLE
9. ZYGOMATICOSCUTULARIS MUSCLE
10. CAUDAL SURFACE OF THE AURICULAR CARTILAGE
11. APEX OF THE AURICULAR CARTILAGE
12. ROSTRAL MARGIN OF THE AURICULAR CARTILAGE
13. CAUDAL MARGIN OF THE AURICULAR CARTILAGE
14. CAVITY OF THE AURICULAR CARTILAGE
15. EXTERNAL CHONDROUS AUDITORY CANAL
16. CERVICOAURICULARIS PROFUNDUS MUSCLE
17. CERVICOAURICULARIS SUPERFICIALIS MUSCLE

SECTION 23: THORACIC LIMB LATERAL ASPECT

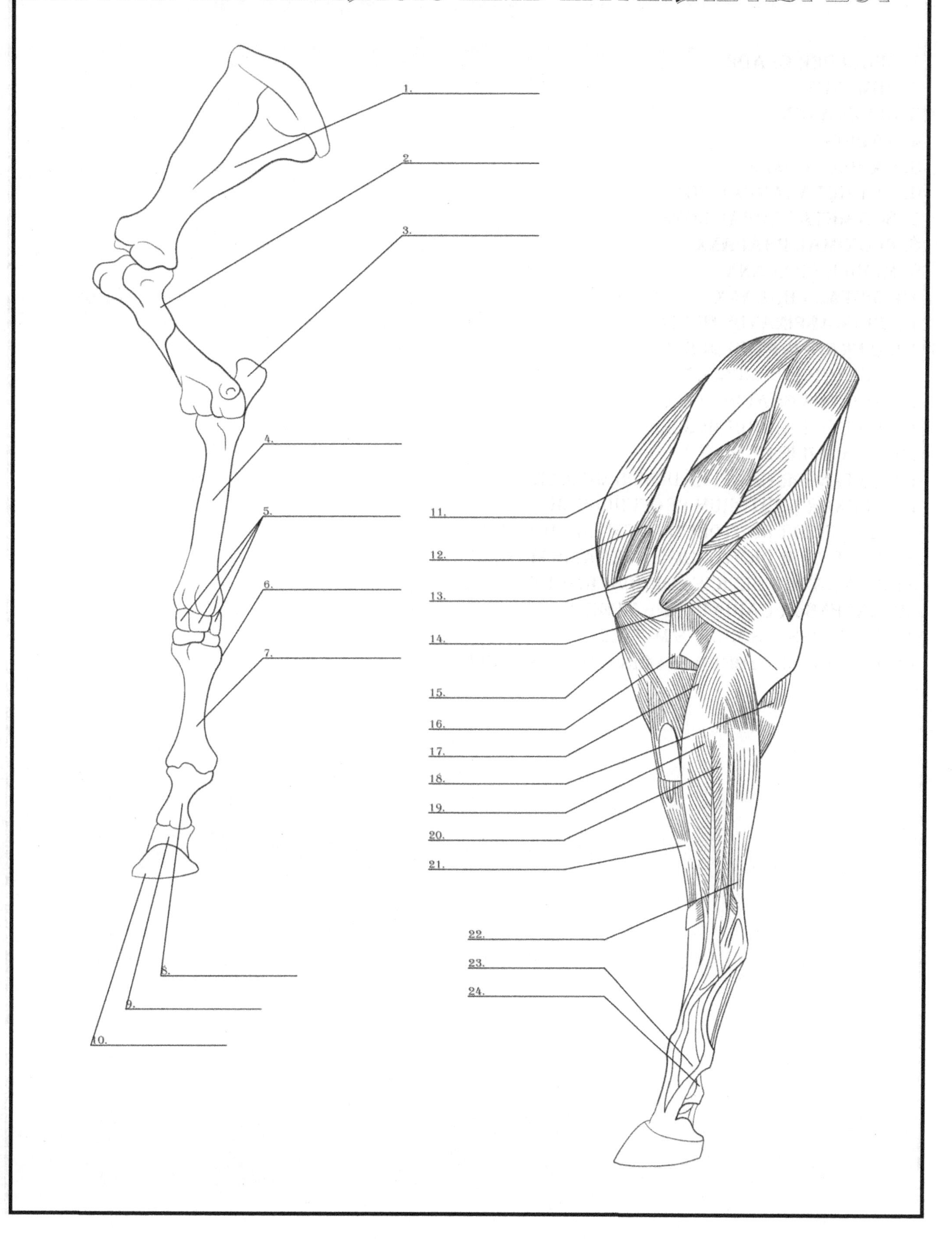

SECTION 23: THROATIC LIMB LATERAL ASPECT

1. SHOULDER BLADE
2. HUMERUS
3. OLECRANON
4. RADIUS
5. CARPAL BONES
6. 4TH METACARPAL BONE
7. 3RD METACARPAL BONE
8. PROXIMAL PHALANX
9. MIDDLE PHALANX
10. DISTAL PHALANX
11. SUPRASPINATUS MUSCLE
12. INFRASPINATUS MUSCLE
13. DELTOIDEUS MUSCLE
14. TRICEPS BRACHII MUSCLE
15. BICEPS BRACHII MUSCLE
16. BRACHILIS MUSCLE
17. EXTENSOR CARPI RADIALIS MUSCLE
18. FLEXOR DIGITORUM PROFUNDUS MUSCLE
19. EXTENSOR DIGITORUM COMMUNIS MUSCLE
20. EXTENSOR DIGITORUM LATERALIS MUSCLE
21. ABDUCTOR POLLICIS LONGUS MUSCLE
22. EXTENSOR CARPI ULNAS MUSCLE
23. INTEROSSEUS MEDIUS MUSCLE
24. FLEXOR DIGITORUM SUPERFICIALIS MUSCLE

SECTION 24: THORACIC LIMB CRANIAL ASPECT

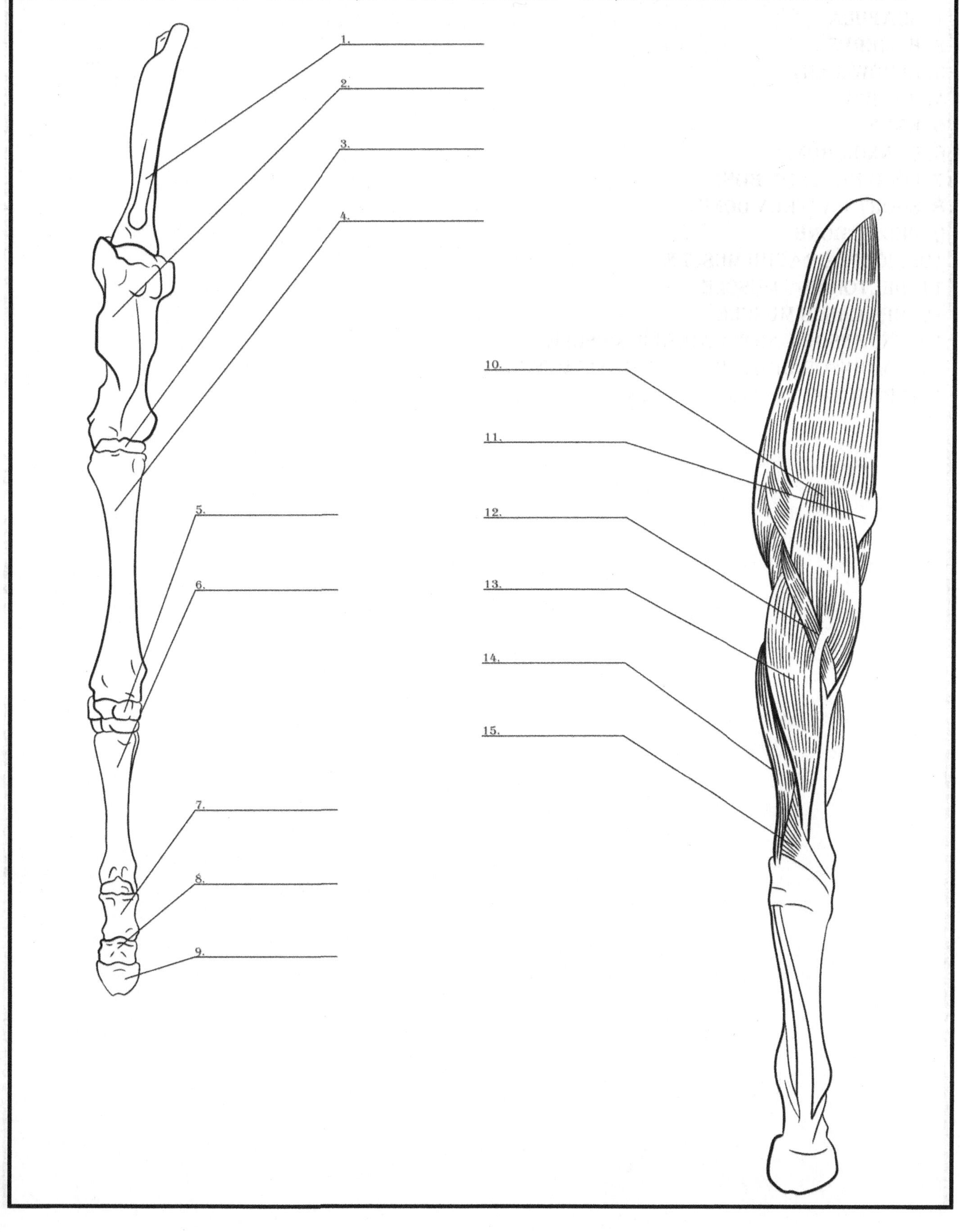

SECTION 24: THROATIC LIMB CRANIAL ASPECT

1. SCAPULA
2. HUMERUS
3. ELBOW JOINT
4. RADIUS
5. KNEE
6. CANNON BONE
7. LONG PASTERN BONE
8. SHORT PASTERN BONE
9. PEDAL BONE
10. BICEPS BRACHII MUSCLE
11. DELTOIDEUS MUSCLE
12. BRACHILIS MUSCLE
13. EXTENSOR CARPI RADIALIS MUSCLE
14. EXTENSOR DIGITORUM COMMUNIS MUSCLE
15.ABDUCTOR POLLICIS LONGUS MUSCLE

SECTION 25: PELVIC LIMB LATERAL ASPECT

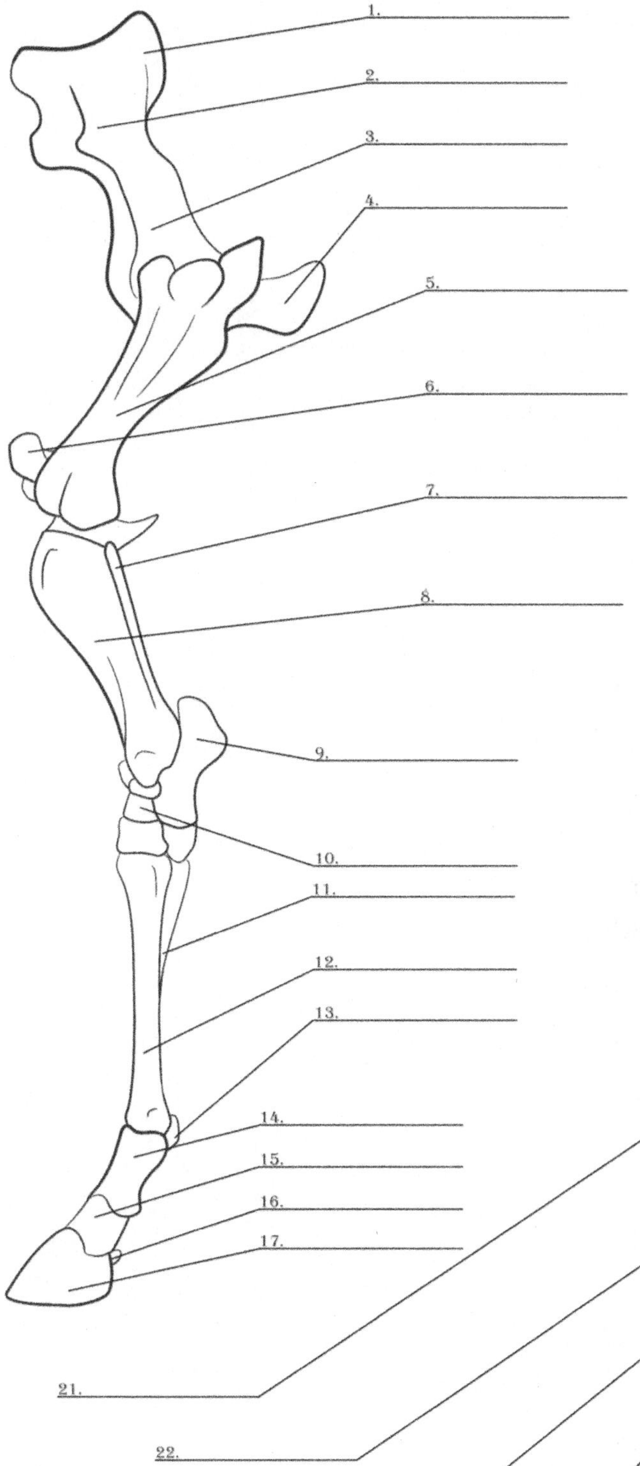

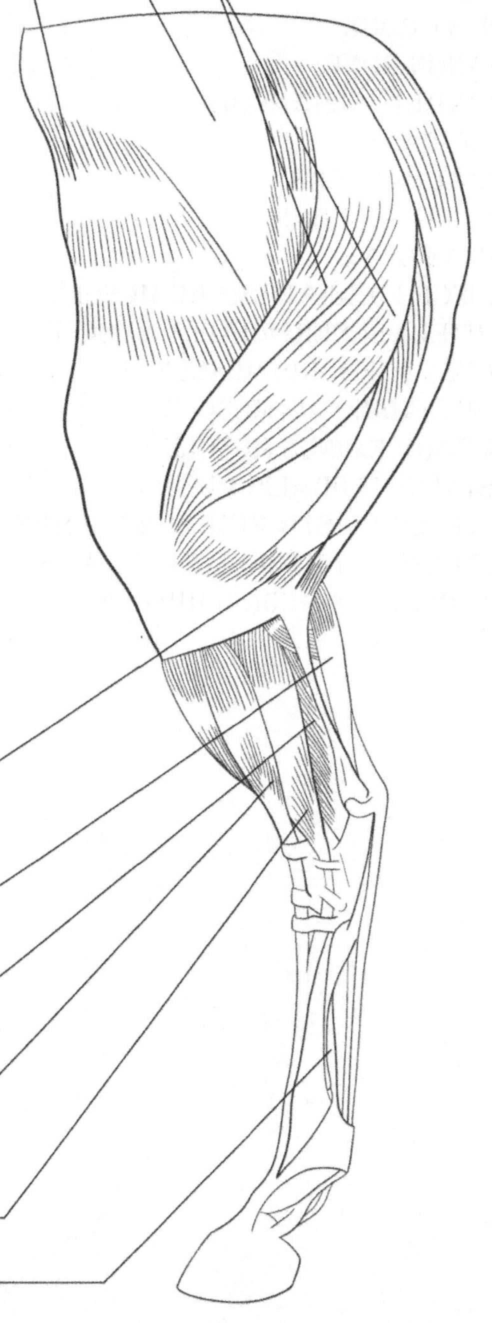

1. ______
2. ______
3. ______
4. ______
5. ______
6. ______
7. ______
8. ______
9. ______
10. ______
11. ______
12. ______
13. ______
14. ______
15. ______
16. ______
17. ______
18. ______
19. ______
20. ______
21. ______
22. ______
23. ______
24. ______
25. ______
26. ______

SECTION 25: PELVIC LIMB LATERAL ASPECT

1. SACRAL TUBEROSITY
2. WING OF ILIUM
3. PELVIS
4. POINT OF BUTTOCK
5. FEMUR
6. PATELLA
7. FIBULA
8. TIBIA
9. CALCANEUS
10. TARSALS
11. SPLINT BONE
12. CANNON BONE
13. PROXIMAL SESAMOIDS
14. LONG PASTERN
15. SHORT PASTERN
16. NAVICULAR BONE
17. COFFIN BONE
18. TENSOR FASCIAE LATAE MUSCLE
19. GLUTEUS SUPERFICIALLY MUSCLE
20. BICEPS FEFMORIS MUSCLE
21. SEMITENDINOSUS MUSCLE
22. GASTROCEMINUS MUSCLE
23. TIBIALIS CAUDALIS MUSCLE
24. EXTENSOR DIGITORUM LONGUS MUSCLE
25. EXTENSOR DIGITORUM LATERALIS MUSCLE
26. INTEROSSEUS MEDIUS MUSCLE

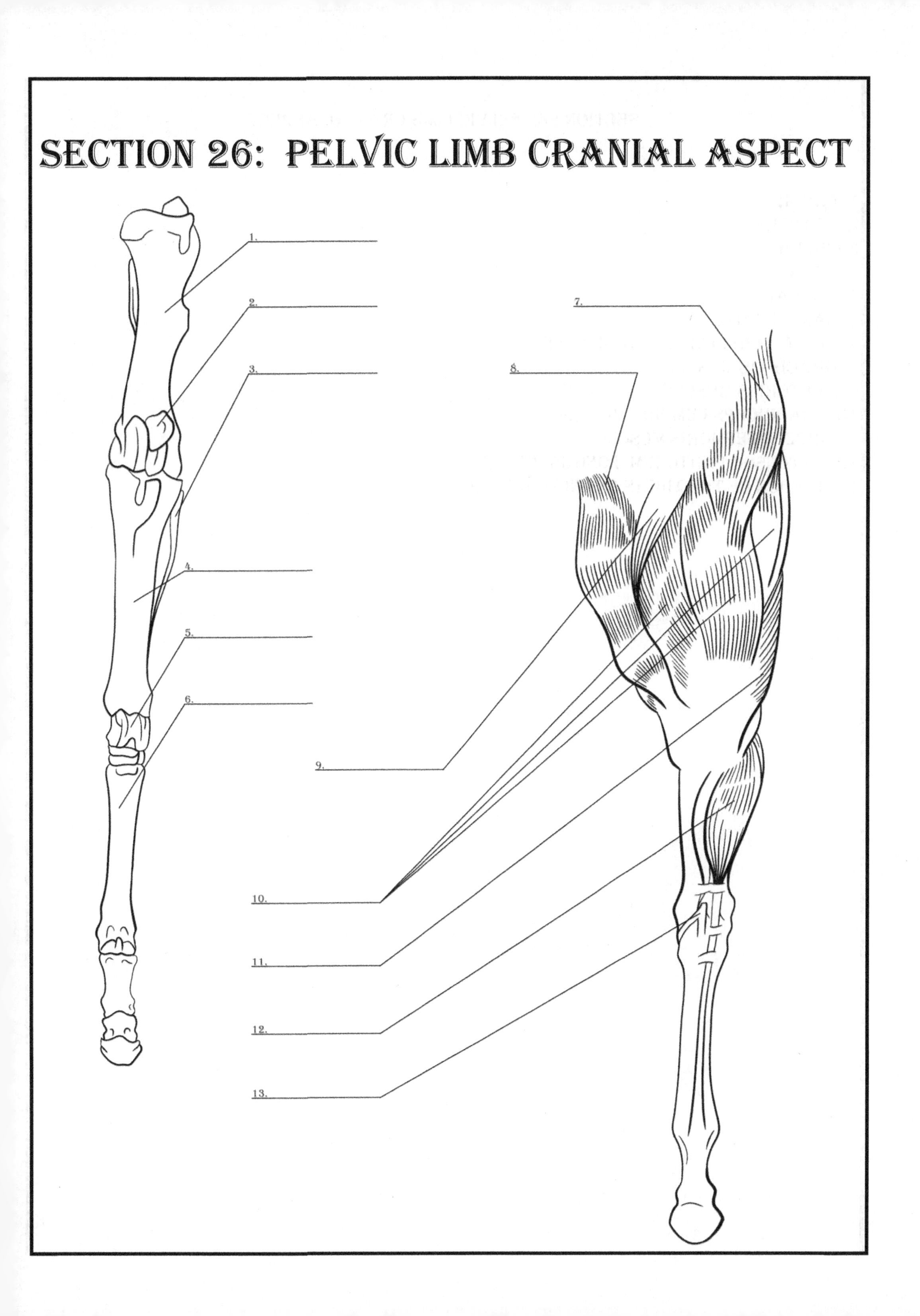
SECTION 26: PELVIC LIMB CRANIAL ASPECT
1.
2.
3.
4.
5.
6.
7.
8.
9.
10.
11.
12.
13.

SECTION 26: PELVIC LIMB CRANIAL ASPECT

1. FEMUR
2. PATELLA
3 FIBULA
4. TIBIA
5. TARSALS
6. CANNON BONE
7. TENSOR FASCIAE LATAE MUSCLE
8. GRACIALIS MUSCLE
9. SARTORIUS MUSCLE
10. QUADRICEPS FEMORIS MUSCLE
11. BICEPS FEFMORIS MUSCLE
12. EXTENSOR DIGITORUM LONGUS MUSCLE
13. TENDON OF THE TIBIAS CRANIAL MUSCLE

SECTION 27: THE HOOF OF THE HORSE 1

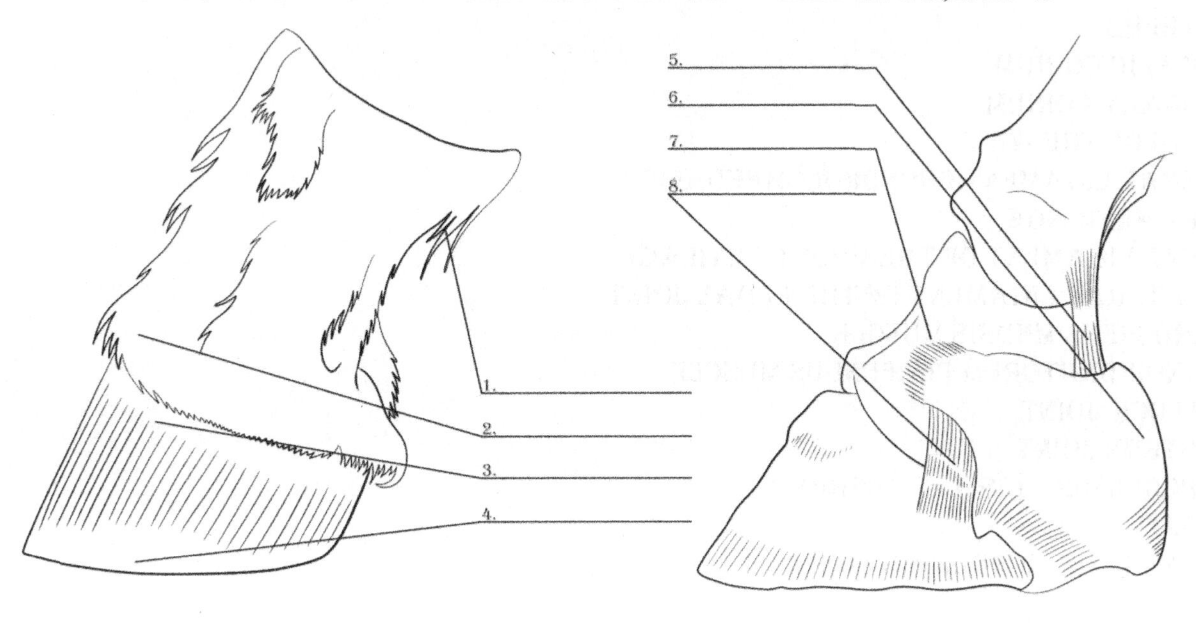

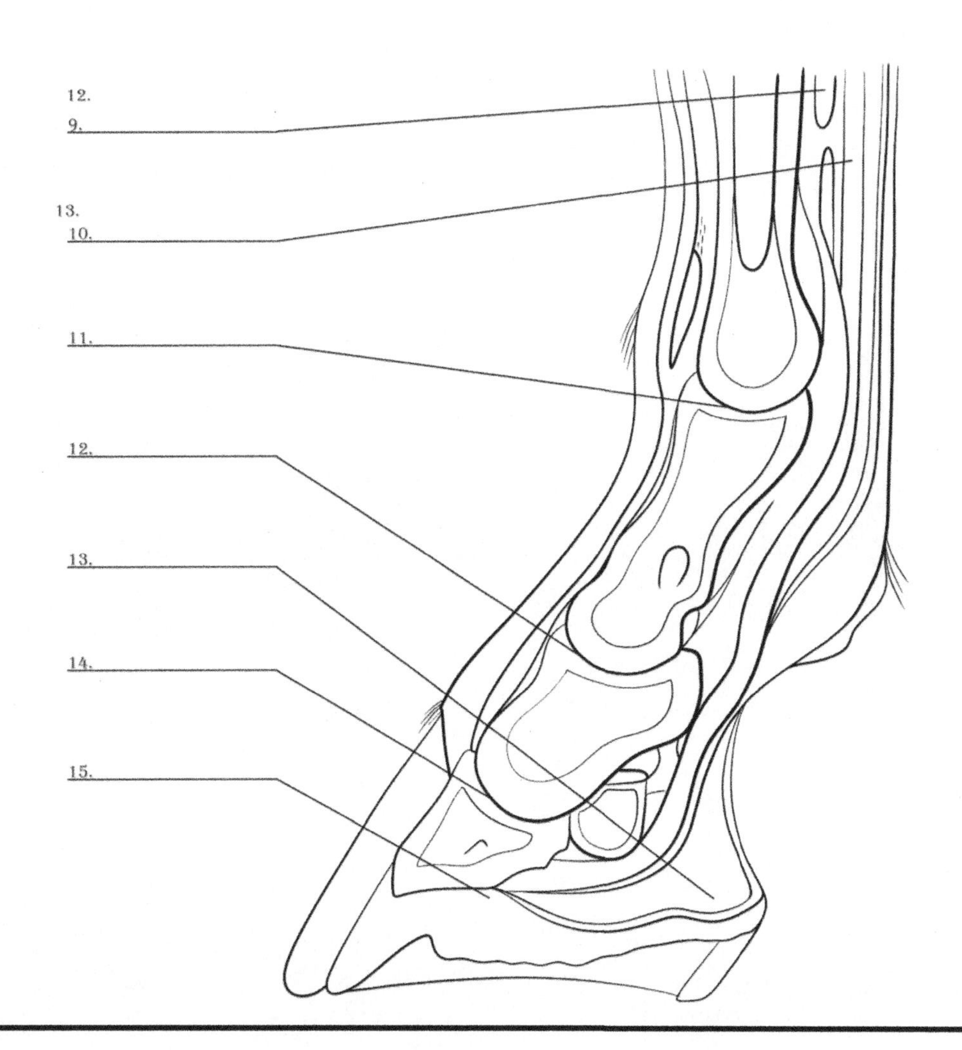

SECTION 27: THE HOOF OF THE HORSE 1

1. FEATHERS
2. PERIOPLIC CORIUM
3. CORONARY CORIUM
4. CORIUM OF THE WALL
5. LATERAL LIGAMENT CHONDROCOMPEDALIS
6. HOOF CARTILAGE
7. DORSAL LIGAMENT OF THE HOOF CARTILAGE
8. COLLATERAL LIGAMENT OF THE PEDAL JOINT
9. INTEROSSEUS MEDIUS MUSCLE
10. FLEXOR DIGITORUM PROFUNDUS MUSCLE
11. FETLOCK JOINT
12. PASTERN JOINT
13. HYPODERMIS (DIGITAL CUSHION)
14. PEDAL JOINT
15. HORNY FROG (EPIDERMIS CUNEI)

SECTION 28: THE HOOF OF THE HORSE 2

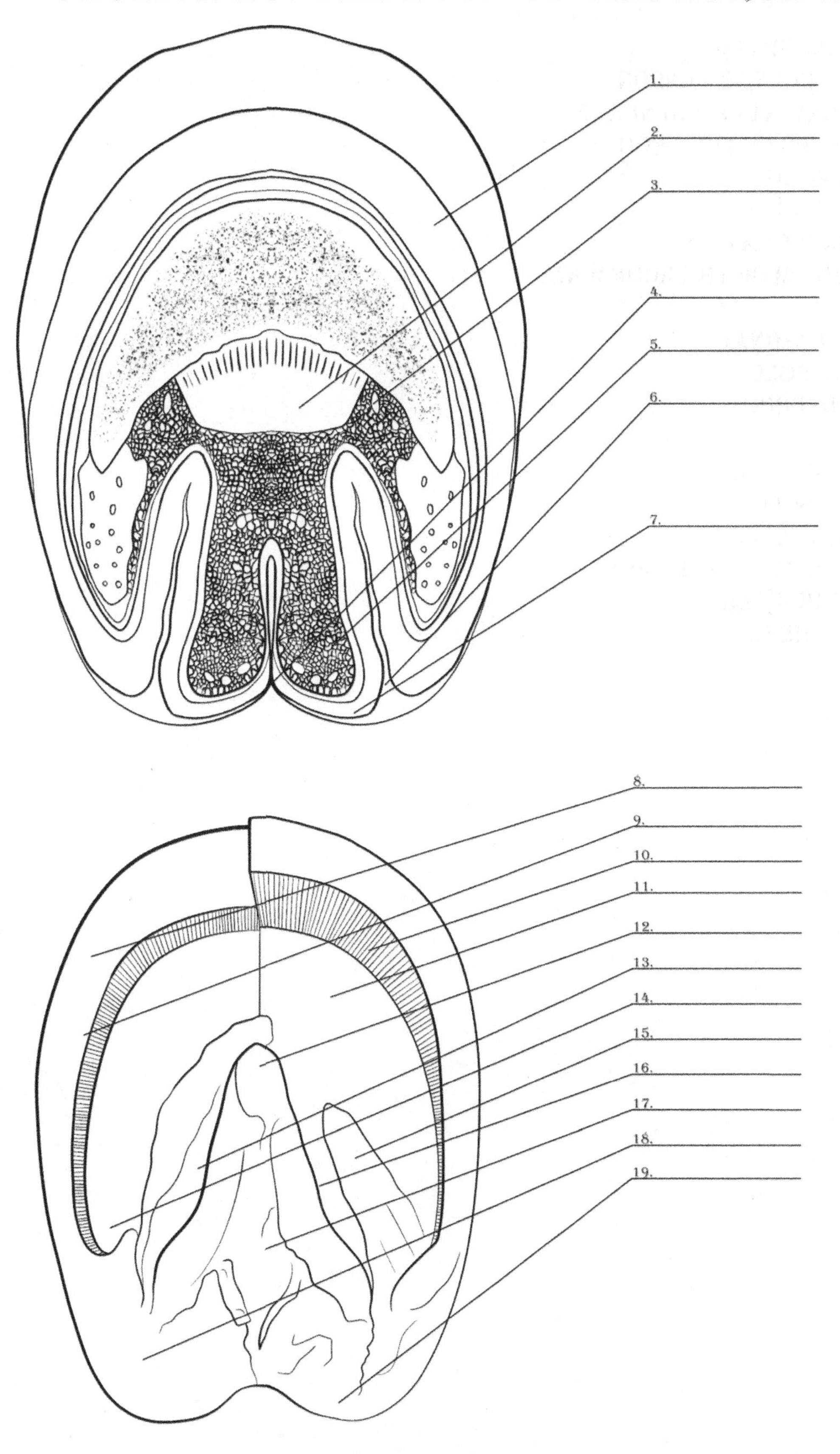

SECTION 28: THE HOOF OF THE HORSE 2

1. CORONARY EPIDERMIS
2. DEEP DIGITAL FLEXOR TENDON
3. MEDIAL ARTERY, VEIN AND NERVE
4. CENTRAL SULCUS OF THE FROG
5. CRUS OF THE FROG
6. PARACUNEAL SULCUS
7. BAR (PARS INFLEXA)
8. STRATUM MEDIUM OF THE HOOF WALL
9. WHITE LINE
10. EPIDERMAL LAMINAE
11. BODY OF THE SOLE
12. APEX OF THE FROG
13. BAR
14. ANGLE OF THE SOLE
15. CRUS OF THE SOLE
16. COLLATERAL SULCUS
17. CENTRAL SULCUS OF THE FROG
18. ANGLE OF THE WALL
19. BULB OF THE HEEL

SECTION 29: THE HEART OF THE HORSE

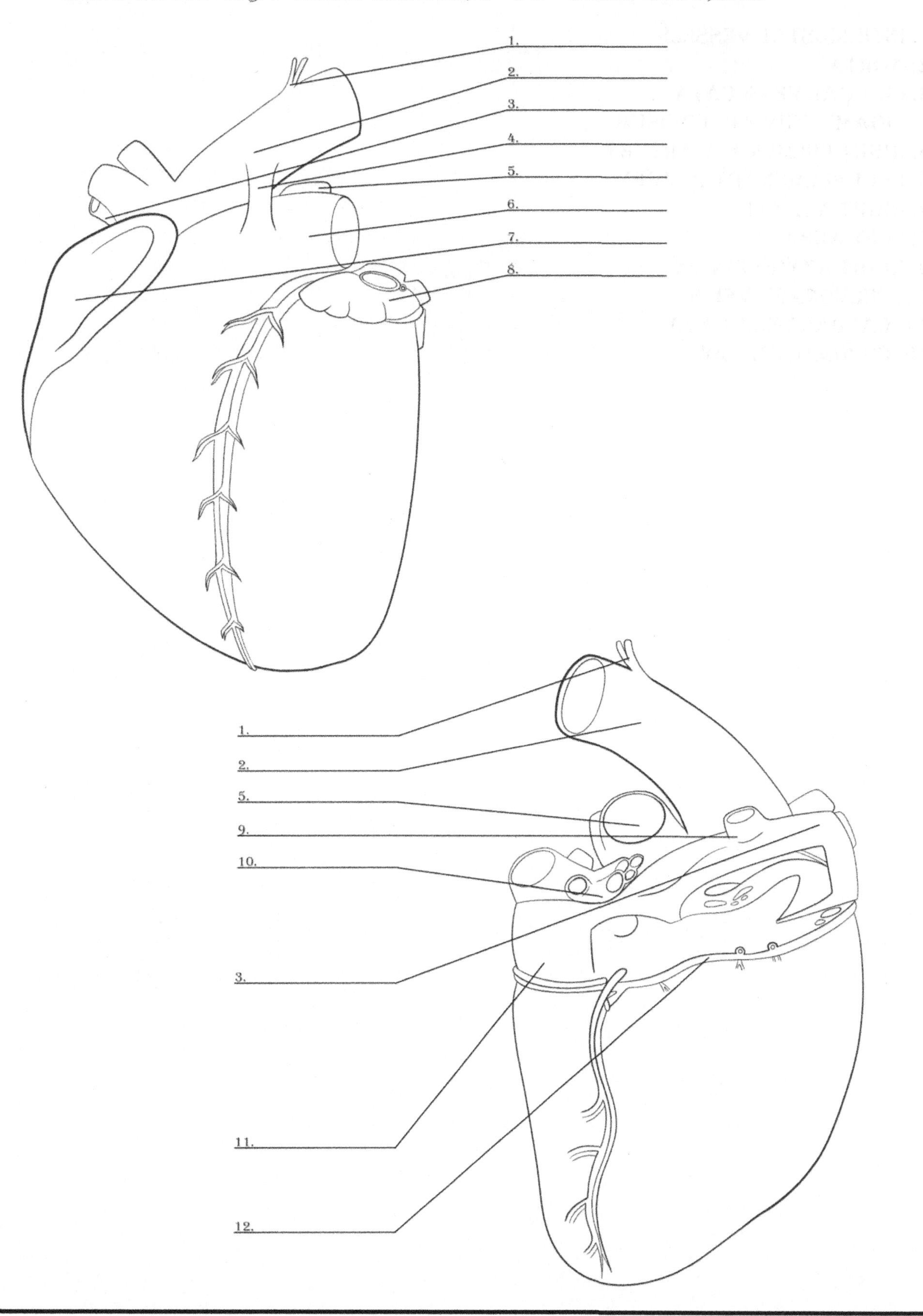

SECTION 29: THE HEART OF THE HORSE

1. INTERCOSTAL VESSELS
2. AORTA
3. CRANIAL VENA CAVA
4. LIGAMENTUM ARTERIOSUM
5. RIGHT PULMONARY ARTERY
6. LEFT PULMONARY ARTERY
7. RIGHT AURICLE
8. LEFT AURICLE
9. RIGHT AZYGOUS VEIN
10. PULMONARY VEINS
11. CAUDAL VENA CAVA
12. CORONARY GROOVE

SECTION 30: THE LUNGS OF THE HORSE

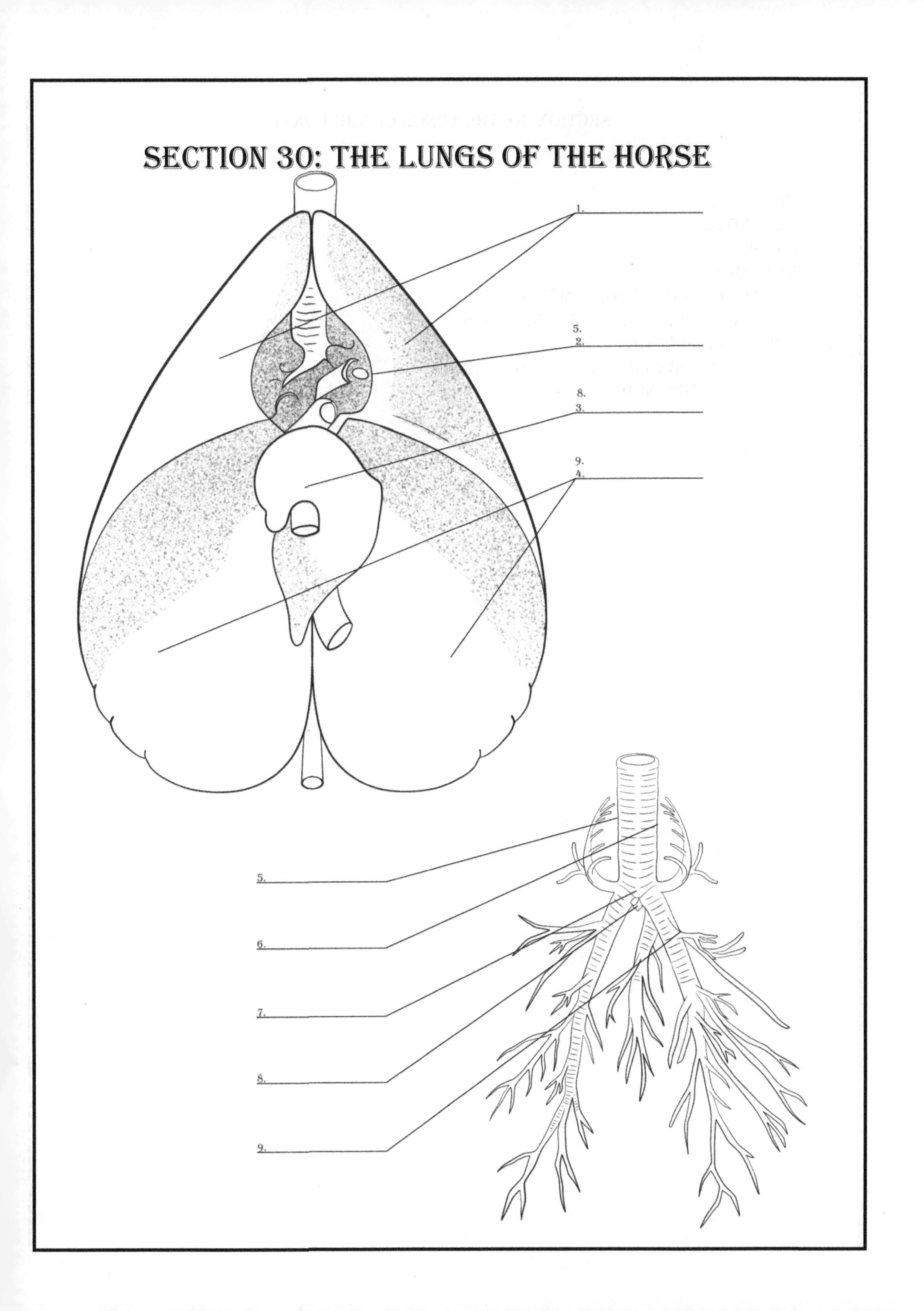

SECTION 30: THE LUNGS OF THE HORSE

1. CRANIAL LOBES
2. CARDIAC NOTE
3. ACCESSORY LOBE
4. CAUDAL LOBES
5. LEFT TRACHEOBRONCHIAL LYMPH NODES
6. RIGHT TRACHEOBRONCHIAL LYMPH NODES
7. TRACHEAL BIFURCATION
8. MIDDLE TRACHEOBRONCHIAL LYMPH NODES
9. PULMONARY LYMPH NODES

SECTION 31: THE SPINAL CORD OF THE HORSE

1.

3.

4.

2.

5.

7.

6.

8.

9.

SECTION 31: THE SPINAL CORD OF THE HORSE

1. LATERAL VERTEBRAL FORAMEN
2. CERVICAL PART
3. INTERVERTEBRAL FORAMEN
4. CERVICAL THICKENING
5. THORATIC PART
6. LUMBAR PART
7. LUMBAR THICKENING
8. SACRAL PART
9. LUMBOSACRAL FORAMEN

Made in the USA
[illegible]
[illegible] December 2021

Made in the USA
Las Vegas, NV
04 December 2024